Multivariate Statistics for the Environmental Sciences

Multivariate Statistics for the Environmental Sciences

Peter J. A. Shaw
School of Life Sciences, University of Surrey Roehampton, UK

John Wiley & Sons, Ltd

First published in Great Britain in 2003 by
Hodder Arnold, a member of the Hodder Headline Group,
338 Euston Road, London NW1 3BH

John Wiley & Sons Ltd, The Atrium, Southern Gate, Chichester, West Sussex, PO19 8SQ, United Kingdom

For details of our global editorial offices, for customer services and for information about how to apply for permission to reuse the copyright material in this book please see our website at www.wiley.com.

ISBN 978-0-470-68923-3

British Library Cataloguing in Publication Data
A catalogue record for this book is available from the British Library

Library of Congress Cataloging-in-Publication Data
A catalog record for this book is available from the Library of Congress

2 3 4 5 6 7 8 9 10

Contents

Preface

Students of environmental subjects operate in a diverse range of academic disciplines. Some concern themselves principally with chemistry, or ecological matters, while others concentrate on the human environment. Most degrees containing the word 'environment' in their title encourage a choice of modules encompassing all aspects of the subject. Despite this diversity, there is one subject which is required by all taught courses: data analysis. This is frequently taught as part of the wider topic of research methods, or may be given alternative titles such as 'information strategies', but an introduction to elementary statistics is universally seen as essential.

This (frequently unpopular) requirement arises from the need to be able to handle and interpret data in order to deal with the morass of noisy and poorly correlated information produced by many environmental investigations. Researchers in pure chemistry or physics can usually expect such high levels of signal to noise in their measurements that visible scatter around a line is a sign of poor technique. By contrast, the majority of experiments or measurements conducted under field conditions generate results which appear at first sight to be random noise and from which trends or differences need to be extracted with suitable statistical procedures. Consequently a prerequisite to interpreting environmental data is a basic understanding of statistics.

Undergraduates are taught concepts such as probability, randomness, null and alternative hypotheses. They go on to learn about association between variables and differences within them, how to calculate some means and standard deviations, t tests and r values. This foundation is vital, and is often called into play when the student undertakes some independent project.

It is usually only once field data have been collected that the limitations of basic statistics courses become apparent. The difficulties arise from the relatively large number of variables that arise from environmental measurements. Any species encountered, each soil property or body length measured, each element assayed or pollutant monitored, each question on a questionnaire becomes a new variable which needs inspection. Instead of the small groups of simple data examined in lectures, a few hours collection of field data could generate hundreds of items of information about many different environmental features. Ecologists might collect data involving many species (upwards of 20 in most habitats), soil scientists or geologists might record five to ten physical measurements for each sample. Sets of data such as these are known as 'multivariate', meaning that each observation is defined by the values of multiple variables. Each of these variables is equally worthy of study – it is not generally possible to know in advance which species will prove to be sensitive to human impact, or which soil property will be most affected by proximity to a roadside. Consequently, any question that may be asked about the dataset must be asked of each variable, one by one.

The problem of analysing multivariate data comes not so much from the number of calculations involved (due to the easy availability of computing power) as in the number of answers obtained. A 30-variable dataset will generate 30 different comparisons (ANOVA, t tests, etc.), and if one explored correlations between variables one would have to inspect 435 different correlations. Given this many tests, one would expect almost 11 significant results from random data, and many more from genuine environmental datasets (where many variables are generally co-correlated). Such a mass of statistical information requires further simplification before it is usable.

The body of techniques which has been developed to help in such situations is known as multivariate analysis. It has been known for many years that the work involved in dealing with multivariate data can be greatly reduced using these analytical tools. The use of multivariate statistics is routine in research papers, but these are rarely taught to undergraduates. In order to familiarise themselves with the unfamiliar world of multivariate statistics, students who turn to textbooks find that there is a wide choice of books – but only if they are familiar with matrix algebra. Most people aren't!

While co-supervising a PhD on lichen responses to air pollutants, I recommended that my student simplify his six physiological measurements using multivariate statistics. He dutifully explored the multivariate statistics books in the university library, and reported the entire selection to be 'daunting'. Exploring the matter further, I was unable to find any book on multivariate statistics that my student could actually understand. In the end I tutored him in the subject myself, and he was able to use a multivariate technique called principal components analysis in his thesis. The results were so neat that I have included them as an example in this book (Example 6.5.3). This showed that students with no pretence at a mathematical background can be taught to understand and apply multivariate statistics, and will benefit from the experience. It also highlighted the dearth of introductory texts on multivariate analyses which bridge the gap between normal undergraduate courses and the mainstream multivariate textbooks. This book is intended to act as that bridge.

A second problem facing the would-be user of multivariate techniques is the bewildering proliferation of names and acronyms. Not only are there a great many multivariate techniques to choose from, but most of them have acquired at least two different names plus a corresponding pair of acronyms. Worse, at least one acronym is sometimes (erroneously) applied to two different techniques. It is hard enough for most environmental students to come to terms with eigenvalues and operations on data matrices, without having to work out that PCO refers to a form of MDS which some packages choose to call CMDS, or that RA means the same technique as CA while CCA is applied to two different techniques offered by the computer package CANOCO. This lack of standardisation can cause serious confusion even to experienced researchers! Consequently this book includes a chapter that attempts to describe, very briefly, the properties, uses, dangers, and synonyms of all the multivariate techniques regularly used by environmental researchers. This is followed by an appendix listing all the standard names of multivariate techniques along with their synonyms, and all the acronyms likely to be encountered, and a glossary. These sections are intended to act as a quick reference when an unfamiliar name is encountered, and will not explain any more than an absolute minimum of underlying theory. The techniques described in detail in Chapters 2–9 should be enough to deal effectively with the great majority of environmental datasets.

Purists may correctly object that diversity indices and multiple linear regression (Chapters 2 and 3) are not multivariate, because they do not produce multivariate output, with multiple orthogonal axes of progressively decreasing importance. I have chosen to include these here because they are tools that a researcher may find helpful in dealing with multivariate datasets. In addition, multiple linear regression forms the basis of later truly multivariate techniques, and has pitfalls that need to be understood.

The detailed chapters (2–9) follow a common format. The chapter starts by explaining the basic concepts underlying each multivariate technique. This is followed by simple worked examples, using small datasets. A set of datasets will be introduced in Chapter 1, and then re-analysed by successive different techniques to demonstrate the operation of each and the outputs they produce. (It is highly instructive to compare the patterns picked up in these familiar datasets by these differing analyses.) Finally, where appropriate, published examples from literature are introduced to show how the technique has been used in practice.

My hope is that this book will open up the usage of multivariate statistics to a wider population than is the case at present, and in doing so improve the quality of analysis given to environmental data. The principles used are sufficiently general that researchers from a wide variety of other disciplines will also find the book to be a useful source of reference.

The book is best read sequentially, since each chapter tends to rely on concepts introduced in previous chapters. However, I hope that the explanations given in each chapter are clear enough to allow dipping into selected chapters as the need arises. Above all, I want this book to be helpful.

Peter Shaw
University of Surrey Roehampton

1

Introduction

1.1 What is meant by multivariate statistics?

People turn to statistics as a tool, to assist them in the interpretation of data. This book is primarily aimed at readers who will be dealing with data collected from environmental sources, probably collected in the field. The range of types of data collected by environmental scientists is very wide. A preliminary list would include vegetation composition (on a percentage cover, presence/absence or other scale), densities of animal species, chemical analyses on air/soil/water/mineral samples, geomorphological measurements and even questionnaire returns. All these will generate large sets of data whose analysis can be problematic and which may potentially benefit from multivariate analyses.

The assistance that statistics can provide takes two distinct forms. One approach is to use statistics to condense data so that it may be described efficiently – boiling down the information to provide a succinct summary. Techniques used for this purpose are known as descriptive statistics. The aim of descriptive statistics is not to test out an idea, but to supply a simplified overview of the information, either as an end in itself or to assist with the formulation of testable hypotheses.

Alternatively, statistics can be used to allow formal testing of hypotheses based on calculations of probability – this is the field known as inferential statistics. The researcher invokes calculations which estimate how likely it is that the observed pattern could occur in random data, and then decides whether or not to accept the possibility that the patterns found are indeed random. As taught by most undergraduate courses the field of descriptive statistics covers calculation of means, standard deviations, interquartile ranges, etc. while inferential statistics includes a large battery of tests including Student's t test, analysis of variance, etc.

However it should be noted that these approaches generally involve analysing just one variable at a time. The technical term for this is univariate. It is impressed on students from school days that one must not add apples to oranges – in other words different variables need to be handled separately. Given data on soil pH and water content, one may calculate a mean pH and a mean water content, but not a mean that includes both pH and water content. (Of course there is nothing to stop you doing such a calculation, except that the answer would be meaningless.) The same applies in the case of univariate inferential statistics. If the soil samples were taken from two different sites one could use a t test to examine whether pH differed between sites, and separately whether water content differed, but one cannot use a t test to examine pH and water at the same time. The mean is a univariate descriptive statistic, a t test is a univariate inferential statistic.

The only exception to this rule which is routinely encountered on elementary statistics course is the situation where two different variables are plotted against each other on a graph. The independence of the two variables is retained because they form the axes of a two-dimensional space. Since two variables are involved this is a bivariate situation, and correlation tests arising out of this are bivariate tests. Calculation of best-fit lines can be thought of as bivariate descriptive statistics, although these calculations tend to be taught alongside correlation analysis (which does supply a probability value so is inferential).

In principle one could provide separate names for statistical approaches to handling three variables simultaneously (trivariate), four variables (quadrivariate), etc. In fact these terms are not used, and instead such techniques are known as multivariate statistics, meaning many variables. Table 1.1 summarises the distinctions between univariate, bivariate and multivariate approaches to data handling.

Thus the term 'multivariate statistics' refers to the extensive collection of tools available to analyse three or more variables at the same time. In fact most multivariate techniques can validly be used in bivariate situations, although it would be unusual to do so. With multivariate statistics you can handle apples, oranges and pears at the same time! To a good first approximation multivariate techniques all come under the heading of descriptive statistics, although in some cases it is possible to generate a testable hypothesis and to estimate probability values (usually by repeatedly imposing a random shuffling on a dataset to measure directly the likelihood of obtaining your actual result from your dataset). The usefulness of multivariate techniques comes from their ability to supply a new form of description, namely the description of patterns between variables. This usually takes the form of diagrams showing the distributions of observations in rather abstract mathematical spaces, which can be off-putting to inexperienced users. The best approach is not to worry too

Table 1.1 An overview of the differences between univariate, bivariate and multivariate statistics.

	Univariate data	Bivariate data	Multivariate data
Number of variables	One	Two	Three or more
Descriptive statistics	Information on the value or range of one measured variable. Examples: mean, standard deviation, skewness, etc.	A numerical function describing the relationship between two independent variables. This is usually given as a best-fit straight line.	A numerical function describing patterns within or relationships between indefinitely many variables. These are usually presented as diagrams showing patterns in abstract mathematical spaces. Most multivariate techniques come under this heading.
Inferential statistics	Calculations of probability based on observed patterns found within one variable. Examples: t test, Mann–Whitney U test, analysis of variance, etc.	Calculations of probability based on observed patterns found between two variables. This usually involves calculation of a bivariate correlation coefficient (Spearman's or Pearson's).	Calculations of probability based on observed patterns found between an indefinite number of variables. Multiple linear correlation supplies exact probabilities, while CCA and related techniques supply probabilities using Monte Carlo techniques.

much about the mathematics needed to derive the picture, but instead to try to interpret the pattern in terms of your intuitive understanding of the data. This often gives deep insights into your results, and allows the formation of testable hypotheses about the system under study.

Multivariate statistics is normally thought of as an aid to hypothesis generation, to be followed up by standard univariate tests, although there is a tendency to regard them as an end in themselves. An illuminating quote about the subject comes as a verbal quip by one of my PhD supervisors Professor Michael Usher: 'Multivariate statistics tell you what you already know.' (with an implicit second section '… but couldn't quite put your finger on'). A competent researcher should have a gut feeling for the major trends within a set of results, but will often be overwhelmed by the volume of data until it is simplified by the application of multivariate tools.

The lack of inferential statistics (i.e. probability levels and hypothesis testing) in most multivariate analysis should not be seen as reducing the value of these techniques, since there are ways in which significance tests may be applied to the output from many of these techniques. This approach is explained in more detail in the relevant chapters.

Although the field of multivariate statistics has its roots in mathematical results from the first half of the twentieth century, their widespread usage only became possible with the advent of accessible computing power in the 1950s and 1960s. Modern researchers may find it hard to credit that prior to this, statisticians would spend days performing the manual calculations needed for one analysis of variance or multiple regression which an undergraduate could now call up off a PC with a few clicks of a mouse. The calculations needed for many of the techniques described in this book are substantially lengthier than those required for univariate analyses, often involving iterative calculations which are time-consuming even for modern PCs, and performing a simple ordination represented days or weeks of manual computation. The arrival of machines such as the Ferranti Pegasus and Mercury computers in the 1950s was seen as a major advance, despite their slowness and need to be programmed in machine code for each calculation (Jeffers, 1995). Remarkably, at least one serious statistician once suggested that one or two computers would suffice for all the statistical calculations likely to be performed in the UK (Swann, 1953). As late as 1958, one of the reasons for the take-up of Bray–Curtis ordination (Chapter 5) was its amenability to computation by hand.

Computer power and availability grew steadily through the 1960s, with most university researchers having access to computer data analysis by the early 1970s. With the realisation that long calculations were no longer any barrier to data analysis, there was a proliferation of techniques developed by statisticians, numerate environmental scientists and others, all seeking to exploit the power of computers to interpret large sets of data. Many of these new techniques were multivariate, since such analyses are inherently demanding of computational resources. This flowering of computer algorithms can be seen as analogous to the Cambrian explosion, which saw a massive explosion in the diversity of life forms. Some techniques never took off, while others were widely used for a short while then declined in importance to the point of extinction. Their stories are given in more detail in Chapter 4 (Section 4.2). A few have survived to become standard features of statistical packages and research publications. Interesting as it would be to explore the obscure and functionally 'extinct' multivariate techniques, the main aim of this book is to introduce the survivors (although Bray–Curtis ordination, a thoroughly 'extinct' technique, is used as a conceptual gateway into ordination in Chapter 5).

1.1.1 Why use multivariate statistics?

Although arcane-sounding, multivariate statistics are widely used by environmental scientists for the good reason that they can save a great deal of analytical work. The need for multivariate statistics can be appreciated by considering that even a small biological survey (e.g. counting abundance of plant species along a transect) will routinely generate datasets with 30 or 40 species and five environmental variables (with more serious work generating correspondingly more data). Each species recorded will require separate statistical analysis. Any of the species recorded may be responding to a measured environmental factor, some other unmeasured factor, or simply exhibit a random distribution. Under such circumstances, even detecting a straightforward trend such as a steady gradient is not simple. One can restrict the work by confining analysis to a few 'important' species – but choosing these species may be a difficult and subjective choice.

Standard univariate lines of analysis (such as correlating each species with each environmental factor) are prone to getting bogged down in a morass of statistical information. It is instructive to examine the case of an ecological study designed to compare densities of plants between two contrasting areas, where the total community is found to contain 30 species (a typical value for student project work). We will take it as read that the surveys were correctly designed with respect to methodology and balanced sampling effort (see Section 1.5). Comparing any given species between the two sites is not difficult – the Mann–Whitney U test or Student's t test could both be appropriate. Since each species could, in principle, be of interest, a full analysis of the data would require 30 different statistical tests, one for each species. The problem then lies in interpreting the (often voluminous!) output generated by these tests. One or two significant values would be expected from any dataset, since statistical significance is usually set at 1 in 20. (This is written as $p < 0.05$, meaning that the probability that the pattern uncovered by your results could have occurred by chance is less than 1/20.) By running 30 tests, one expects 1.5 ($=30 \times 1/20$) of them to be significant even when applied to random data. Very likely the analysis would reveal that more than two species differed significantly between the two sites, with different species showing contrasting distribution patterns. Instead of one clear answer, the researcher ends up with 30 different answers, and still needs to find a way of drawing these together into a coherent conclusion. If the question switches to one of association – 'Are there species which selectively occur together or which repel each other?' – the researcher finds that a 30-species dataset generates 435 possible correlations to examine. (This figure is found as follows: Species 1 can be correlated with species 2, or 3, or 4, ... or 30, making a total of 29 correlations. Species 2 has already been correlated with species 1, so only generates 28 correlations. Species 3 generates 27 correlations, ... and species 29 generates 1 correlation. The total number of correlations needed to explore relationships within a 30-species community is $29 + 28 + 27 + \cdots + 1 = 435$.) In fact 30 variables is modest for an ecological dataset. Species-rich communities can generate datasets hundreds of columns wide, and the task of finding meaningful patterns in them is correspondingly greater.

By contrast, the same 30+ variables could be handled as one block of data by multivariate techniques, with a corresponding reduction in workload and in the danger of misinterpreting random noise. Depending on the questions asked about the data, a number of multivariate approaches can be taken. An index of diversity (measuring the abundance of species and the balance of community composition) could be calculated for each community, to see whether it changed in any systematic way. The 30 species could be replaced by a much smaller set (typically 2–4) of newly created, independent variables which would summarise the community composition, and could be subjected to conventional methods of analysis. This process of simplification is known as ordination. All dependent variables could be input to the same regression against one environmental variable – this is multiple regression – or could be related to a second multivariate dataset *en masse*, using canonical correspondence analysis (CCA). Finally, one could search for clustering in the data that suggest the occurrence of distinct communities – the realm of cluster analysis.

These five approaches – diversity indices, multiple regression, ordination, cluster analysis and canonical correspondence analysis – are collectively the subjects of the main chapters of this book. Technically, diversity indices and multiple linear regression are not mathematically multivariate (since they only produce one number as an output, while true multivariate techniques produce a large number in decreasing order of importance). They are included here because they are useful techniques in their own right and because they give the inexperienced student an introduction to the concept of combining variables together to generate new, abstract indices.

To summarise this section, multivariate statistics should be used to explore and describe datasets containing many variables. They allow the generation of hypotheses and suggest patterns to be found within data while requiring relatively little work (at least for the human operator), but only rarely are they an end in themselves.

1.2 The scope of this book

There is now a wide choice of texts which explain the basics of statistical analysis, ranging in scope from simple introductions to probability and data collection upwards to university level texts. It would be invidious to select any particular texts, due to the range of levels available. By contrast, although there are many books about multivariate statistics, few can be described as introductory. The entire subject has a reputation amongst non-users of being impenetrable and incomprehensibly mathematical. This is compounded by the widespread duplication of names and acronyms, with as many as four different names being applied to the same analytical technique according to which textbook/manual is being used.

This is a great pity, and is a serious limitation for the analysis of many datasets. For most students and researchers, the psychological barrier is greater than the problem of software. Most statistical packages give ordination, multiple regression and clustering as standard options, but make little or no attempt in their documentation to explain the rationale behind the techniques nor suggestions as to interpretation of output. (Worse, software manuals rarely warn of the dangers of misinterpretation inherent in many multivariate techniques – a particular problem for cluster analysis and all techniques incorporating multiple linear regression.) The intention of this book is to give naive users a grounding in the fundamentals of multivariate analysis, in order that they should be able to benefit from the application of these techniques to their data.

This book is intended to assist the student involved in the collection and analysis of environmental data. It assumes a knowledge of simple inferential statistics, notably bivariate linear correlation and one-way analysis of variance (One-way ANOVA). Given this basis, the book assumes no knowledge whatsoever of multivariate statistics, and sets out to introduce the basic concepts needed to handle multiple variables in one analysis. The commoner and more useful techniques will be described and their application explained. Appropriate lines of analysis will be explored for some typical environmental datasets.

Among the common techniques, Chapter 5 includes an old ordination technique known as Bray–Curtis ordination, which is rarely met with either in software packages or in research publications. This inclusion is deliberate, since this archaic technique serves as an excellent introduction to many of the most important concepts in multivariate analysis while being simple enough to perform by hand. It is worth understanding the Bray–Curtis procedure even if one never intends to apply it, since it gives insight into the concept of ordination and is similar to the commonly used suite of procedures collectively known as cluster analysis (Chapter 8).

I do not intend to derive the mathematics underlying any of the techniques, nor do I expect that reading this book will give a full understanding of the entire suite of multivariate techniques. For readers requiring more detail, there are many good books which explain the techniques with greater mathematical rigour, listed under the 'Suggested further reading' section (Appendix 4).

The principal aim of the book is to explain the application of the commoner multivariate techniques using examples. The general structure of each chapter will be the same. First the theory of each technique will be introduced, using visual rather than mathematical explanations as far as is possible. Next the technique will be applied to small, model datasets in order to explain the procedures. Finally actual examples from published literature will be given, showing how multivariate techniques can contribute to environmental research.

Chapter 10 follows a different format, in that it aims to give a brief explanation of other multivariate techniques without the depth of explanation given in preceeding chapters. This is followed by an appendix of the synonyms and acronyms of multivariate techniques (itself a fertile source of confusion even for experienced researchers).

1.3 When multivariate statistics might be used

This section will give examples of environmental datasets lifted from published literature, and briefly outline how multivariate analysis facilitated their analysis. At this stage the techniques used will be unfamiliar, so all that will be given will be a brief overview of the nature of the research and the type of analysis used.

It would be useful to start by defining situations where multivariate analysis is *not* needed. In these situations there is just one specific dependent variable, which is incorporated into a designed experiment or is the subject of a survey of its behaviour in response to one well-defined factor. Environmental datasets where no multivariate analysis would be applicable might include measurements of the deposition of *one* pollutant (lead, fluoride, etc.) along a transect away from a pollution source, or agricultural experiments studying biomass yield in response to fertiliser application. Medical trials measuring one factor (blood pressure, survival rates, etc.) in response to application of treatments would also have no need for multivariate analysis, but even here there might be a plethora of background data about the patients' health that could be simplified by multivariate analyses and included as additional information.

The list of situations where analysis would definitely benefit from multivariate statistics is rather longer – in fact this covers any situation where several dependent variables, or several environmental factors, are measured together. A list of particularly common environmental research situations is given below, but this is in no way exhaustive.

- You have species lists from a biological community and want to study the response of the overall community to some factor(s). The community could equally be plants, animals, fungi, or other taxa.
- You have taken multiple measurements to define size/shape of individuals (for taxonomic purposes, to analyse sexual dimorphism, to investigate the impact of an environmental factor on body size, etc.). This would typically cover biological morphometrics, but exactly the same approach could be used for the body dimensions of cars, or the geomorphological features of valleys.
- You have defined samples by a list of chemical data (commonly water composition, air pollutants or soil nutrients) or by biochemical measurements. Instead of analysing each chemical separately it may be more informative and economical to handle them all simultaneously using multivariate techniques.
- You have data on the frequency of biochemical/genetic markers in a set of organisms and wish to explore whether there is evidence of genetically isolated subpopulations within the species.

1.4 Minimum computing requirements

With the exception of diversity indices and Bray–Curtis ordination, all the techniques described in this book require a computer. It is unlikely that computing power will be a limiting factor, unless you are dealing with unusually large datasets. The actual minimum computing power will depend on the software package used. At time of writing most serious packages offering multivariate statistics require an IBM-compatible personal computer running a Windows operating system (Windows 95 or higher). Within a few years this specification will look archaic, and one intention of this book is that it will continue to be useful and relevant despite the inevitable upgrading that will occur in software and computing power.

It might be useful to set this progressive upgrading of computer power into context with a homely example. In 1986 I ran principal components analysis (an ordination technique described in Chapter 6) on a home computer containing 32 kilobytes of RAM, using a program I wrote myself in the computer language BASIC. The computer contained just half the memory capacity of a cheap 1997 digital diary, and BASIC is such a slow and bug-prone language that no serious programmer would use it now. Despite all this, the program worked! In other words, the factors limiting the use of multivariate statistics are understanding of the techniques and the availability of software – not computing power.

1.5 Preparing the data

This book is not about data collection techniques. The practicalities of field work vary between disciplines, and excellent textbooks already exist covering all aspects of environmental fieldwork. Consequently this book will make no attempt to offer guidance on the actual collection of data. However there are important general points to be considered when preparing data for analysis, some of which have a bearing on the conduct of field surveys.

The points to be considered come under four headings:

1. Types of data to collect.
2. Avoiding pseudoreplication
3. Organisation of the data matrix.
4. Preliminary inspection of the data.

1.5.1 Types of data to collect

Information can be gathered in many different ways, not all of which are suitable for all types of analysis. There are four different types of data, given here in order of increasing information content.

Nominal data

These data involve description of categories, where no attempt can be made to put the categories in order. Flowers, soils, etc. may be coded by colour – clearly red is different from brown, but one cannot ask whether red is more or less than brown. Other examples of nominal data that might be entered into environmental datasets could include soil texture, rock type or land use.

Ordinal data

Ordinal data involve categories than can be put into rank order, but where it would not be meaningful to ask for the numerical difference between categories. Animals could be coded as young/immature/adult, or a questionnaire score could range from 1 (very happy)–5 (very unhappy). In both cases the results can be ordered; an adult is older than a youngster and a response of 5 is less happy than a response of 1. However it is not meaningful to ask what the difference in scores means, still less to obtain a ratio of adult divided by youngster! A common and useful example of an ordinal scoring system is the DAFOR scale for recording vegetation cover, a five-point scale where D = dominant, A = abundant, F = frequent, O = occasional, and R = rare.

Continuous data

This is the commonest and most useful type of data, and covers any measurement where the result can (in theory) take any numeric value within some defined range. Continuous data are subdivided into two further categories: **interval** and **ratio** data.

Interval data This class involves measurements which can be set into rank order and for which it is meaningful to calculate the difference between two data points. The only distinction between this class and ratio data concerns the value given to zero; for interval data the zero value is arbitrary, and in consequence it is not meaningful to calculate the ratio of two values. The usual example of such data is temperature; a temperature of 15 degrees is not half a temperature of 30 degrees.

Ratio data This is the highest quality of data, and involves measurements for which rank order, differences and ratios are all meaningful. This is the most important category of environmental data and includes chemical determinations, species counts, flow rates, etc.

Continuous data are preferable for the majority of multivariate techniques. Bray–Curtis ordination (Chapter 5) and cluster analysis (Chapter 8) can safely be used on nominal or

Table 1.2 A sample set of soil measurements, identifying the data types.

Location date depth	*These are metadata: information identifying the origin of the samples.*
Colour (from a colour chart) Texture (loam, clay, etc.)	*Nominal data: these define named classes but cannot be ranked into order.*
pH % organic matter Nitrogen mg g^{-1}	*Continuous data: these may be ranked into order or entered into calculations.*

ordinal data, providing a suitable similarity index is used. Other multivariate techniques will run on non-continuous data, but the results should be treated with caution. While planning fieldwork, it is important to examine the types of data that will be generated and consider whether they will be suitable for your proposed analyses.

Table 1.2 shows a sample dataset of soil properties identifying the measurements' data type.

1.5.2 Avoiding pseudoreplication

Pseudoreplication is a serious statistical trap which has caught many researchers. It is explained fully by Hurlbert (1984), and only the part of the problem relating to field sampling will be covered here. The formal definition of pseudoreplication is the testing of a hypothesis with an inappropriate number of degrees of freedom. In simpler terms, this amounts to organising your data in such a way as to pretend that you have made more independent observations than is actually the case. This pretence is almost always entirely unintended by the researcher and cannot be detected by computer packages (which simply analyse the data they are given), but renders subsequent statistical analyses invalid. Numerous papers and PhD theses have been referred because of accidental pseudoreplication.

In the context of the kind of environmental datasets relevant to this book, a likely cause of pseudoreplication involves combining different types of data that were not collected together. An example is given by Matthew *et al.* (1988), who collected data on the vegetation of seven New Zealand pastures in relation to soil properties. Plant data were collected from 10 randomly placed quadrats per pasture, but soil data were only collected from one point. Thus although the plant and soil data came from the same sites at the same time, the two datasets were not actually collected together.

It would be tempting and straightforward to give the same value for each soil parameter to each quadrat (see Figure 1.1 and Table 1.3). This would be formally invalid, as it is tantamount to claiming that an independent soil sample was taken and analysed for each quadrat, which was not the case. Instead, when the data were analysed by Matthew *et al.* (1994) the correct approach was taken, which was to treat one pasture as the independent unit of observation, so combining mean soil data with mean vegetation data. Note that this meant that a great deal of within-pasture information was lost; only one line of data was entered for each pasture instead of ten. A good rule is that one line of data should contain only measurements taken together in the same place at the same time. An alternative sampling design which would bypass this pseudoreplication problem is given in Figure 1.2, in which each vegetation sample is matched with a soil sample taken from the same spot at

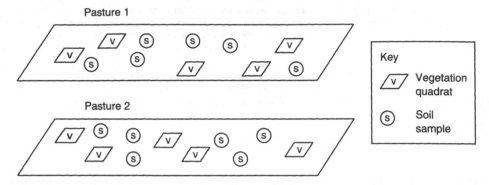

Figure 1.1 A sampling design that is liable to generate the error of pseudoreplication. In this case soil data and vegetation data must be aggregated down to one data point per pasture before they may be pooled together.

the same time. (Note however that if the work were studying the effect of a treatment added to entire pastures, such as a fertiliser or pesticide application, then again consideration of individual quadrats would be pseudoreplication for any treatment applied to a whole pasture, and in this case analyses should instead rely on data aggregated at the level of pasture.)

An example in which pseudoreplication is introduced into a design by constraints is given as Example 3.6.5 (Chapter 3). This also introduces a technique for using the whole of a pseudoreplicated dataset while working within correct degrees of freedom known as jack-knifing (Efron, 1979). There is a great deal more to the issue of pseudoreplication than this; seedlings in the same plant pot should not be regarded as independent, nor should observations in a time series. For a clear but detailed review of this topic the reader is referred to Hurlbert (1984).

1.5.3 Organisation of the data matrix

In order to undertake any effective analysis of any environmental dataset, data need to be organised into a suitable format. The format of the data refers to the shape and layout used to contain the information, and needs careful attention for the simple reason that computers are utterly stupid! Humans can accommodate for irregularities in layout – computers generally cannot, and many analyses have been rendered valueless because of a computer package doing what it was told to do instead of what the operator wanted it to do.

There are a variety of specialised data formats used by different packages, but the simplest, safest and most widely used standard is to store the information in a matrix. A matrix is simply a rectangular array of data (usually numbers, but in this context also alphanumeric characters). If the data are seen as a series of rows, each row is one observation and all rows contain the same number of data items. If the data are viewed as a series of columns, all the numbers in any given column should represent sequential observations of the same variable. If the first number in a column is a pH value, all the numbers in that column should be pH values. Note that the number of columns multiplied by the number of rows equals the number of data points in the matrix, for the same reason that the area of a rectangle is the product of the lengths of its longer side and its shorter side.

Table 1.3 An illustration of how incorrect pooling of field data can generate the error of pseudoreplication. Data are in percentage cover, and the sampling protocol is given in Figure 1.1. In this instance vegetation and soil samples were collected from different locations, so may not validly be pooled except at the level of the pasture.

		Vegetation data		
Pasture	quadrat	*Lolium perenne*	*Festuca ovina*	*Trifolium repens*
1	1	70	40	0
1	2	45	60	5
1	3	75	40	1
1	4	95	10	0
1	5	100	0	0
2	1	40	60	20
2	2	20	50	45
2	3	15	40	60
2	4	25	0	80
2	5	10	20	70

Soil data

Pasture	N mg $100\,g^{-1}$	pH	% organic matter
1	126	6.5	4.2
2	35	6.3	1.3

These two datasets can be merged in two ways, given below. The first approach implicitly states that one soil sample was taken in each quadrat, and is therefore invalid. The resulting inflation in the apparent size of the data collected is an example of the statistical fault known as pseudoreplication. The second dataset merges information at the level of the pasture (using a mean value for each parameter), and is formally valid. It will be seen that this approach loses information about individual quadrats. Abbreviations: Lp – *Lolium perenne*, Fo – *Festuca ovina*, Tr – *Trifolium repens*, N – nitrogen, mg $100\,g^{-1}$, OM – percentage organic matter.

		Pooled data – pseudoreplicated					
Pasture	quadrat	Lp	Fo	Tr	N	pH	Om
1	1	70	40	0	126	6.5	4.2
1	2	45	60	5	126	6.5	4.2
1	3	75	40	1	126	6.5	4.2
1	4	95	10	0	126	6.5	4.2
1	5	100	0	0	126	6.5	4.2
2	1	40	60	20	35	6.3	1.3
2	2	20	50	45	35	6.3	1.3
2	3	15	40	60	35	6.3	1.3
2	4	25	0	80	35	6.3	1.3
2	5	10	20	70	35	6.3	1.3

	Pooled data – Correct approach				
Pasture	Lp	Fo	Tr	N	pH	Om
1	77	30	1.2	126	6.5	4.2
2	22	34	55	35	6.3	1.3

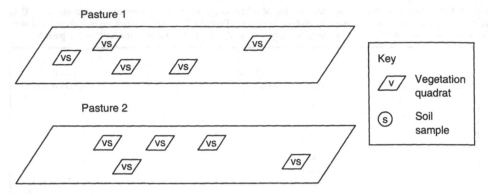

Figure 1.2 This sampling design allows soil data and vegetation data to be analysed together without aggregating data or risking pseudoreplication, since each soil sample is paired with a corresponding vegetation sample.

Table 1.4 A simple data matrix with 13 columns and eight rows. These give the soil properties and percentage cover of plant species on two adjacent locations on Wimbledon common, a sandy lowland heath (site 1), and adjoining spoil mounds of calcareous clay (site 2). Notice that the matrix starts with two columns of metadata – information about the data (in this case site number and quadrat replicate number). Abbreviations: Ae – *Arrhenatherum elatius*, Am – *Achillea millefolia*, cond – electrical conductivity of soil solution (mS cm^{-1}), Cv – *Calluna vulgaris*, Df – *Deschampsia flexuosa*, H_2O – percentage water content of soil after drying to 105°C, Fr – *Festuca rubra*, Hs – *Heracleum sphondylium*, OM – percentage organic matter composition of soil, rep – quadrat replicate number, Site – site number (1 = heathland, 2 = clay spoil), Tr – *Trifolium repens*, Vs – *Vicia sativa*.

Metadata		Soil measurements				Vegetation cover							
Site	rep	pH	cond	OM	H_2O	Am	Ae	Fr	Cv	Df	Hs	Tr	Vs
1	1	4.5	55	26	17	0	0	80	5	25	0	0	0
1	1	5.4	60	16	21	0	0	10	95	0	0	0	0
1	3	5.1	49	34	18	0	0	70	90	30	0	0	0
1	4	4.8	55	27	18	0	0	95	50	5	0	0	0
2	1	7.6	155	5	25	10	70	0	0	0	0	5	1
2	2	7.8	124	7	35	5	85	0	0	0	0	10	0
2	3	7.2	141	6	32	2	60	0	0	0	25	1	5
2	4	7.3	166	8	29	1	50	0	0	0	40	0	3
↑	↑	↑	↑	↑	↑	↑	↑	↑	↑	↑	↑	↑	↑

Thirteen columns

Table 1.4 gives a simple example of a data matrix (which will be used again throughout this book). This consists of soil and vegetation data from Wimbledon Common. The first two columns define the site number and replicate number from which each sample was taken. These are an example of metadata, and are essential to identify the origin of samples but must not be entered into any preliminary analyses, although they may be used later to define groupings in tests such as ANOVA. (It should also be noted that this table includes column headings, which are useful to improve intelligibility but will generally need to be removed before the numbers can be read into a package.)

Assembling such a rectangular data matrix is generally not difficult, though the exact method you will use will depend on your choice of software. One approach might be to use a text editor to create a file containing your data, which can then be read in by a statistics package. This approach has the virtue that, once created, the data file will be in ASCII

Table 1.5 A sample ecological community dataset: this is a subset of the Liphook pine forest fungal data set (listed in full in Table 1.9), listed as one might record observations in a field notebook.

year: 86
Plot: 1
Laccaria proxima: 280
Paxillus involutus: 1

year: 87
Plot: 1
Gomphidius roseus: 1
Laccaria proxima: 1025
Suillus variegatus: 2
Suillus bovinus: 5
Paxillus involutus: 1

year: 88
Plot: 1
Cortinarius semisanguineus: 11
Gomphidius roseus: 7
Inocybe lacera: 15
Laccaria proxima: 1265
Suillus luteus: 3
Suillus variegatus: 102
Suillus bovinus: 299

year: 89
Plot: 1
Cortinarius semisanguineus: 44
Gomphidius roseus: 8
Inocybe lacera: 1
Laccaria proxima: 19
Suillus luteus: 10
Suillus variegatus: 295
Suillus bovinus: 568

year: 90
Plot: 1
Boletus ferrugineus: 1
Cortinarius semisanguineus: 117
Gomphidius roseus: 4
Inocybe lacera: 6
Laccaria proxima: 24
Lactarius rufus: 24
Suillus luteus: 14
Suillus variegatus: 116
Suillus bovinus: 36
Paxillus involutus: 2

This dataset can be reformatted into a sparse matrix in which each column represents counts of a different species, thus:

year	plot	Bf	Cs	Gr	Il	Lp	Lr	Pi	Sb	Sl	Sv
86	1	0	0	0	0	280	0	1	0	0	0
87	1	0	0	1	0	1025	0	1	5	0	2
88	1	0	11	7	15	1265	0	0	299	3	102
89	1	0	44	8	1	19	0	0	568	10	295
90	1	1	117	4	6	24	24	2	36	14	116

Abbreviations: Bf – *Boletus ferrugineus*, Cs – *Cortinarius semisanguineus*, Gr – *Gomphidius roseus*, Il – *Inocybe lacera*, Lp – *Laccaria proxima*, Lr – *Lactarius rufus*, Pi – *Paxillus involutus*, Sb – *Suillus bovinus*, Sl – *Suillus luteus*, Sv – *Suillus variegatus*.

format and can be read into almost any application. If an ASCII file is to be created, it is good practice to ensure that all columns line up, even though the days of needing to specify exact start/end points for each column in a data file have mercifully passed. The advantage of neatly aligning numbers within columns is that erroneous or missing points can be spotted far more easily in a regular layout than in a higgledy-piggledy mess.

A good alternative is to store data within a spreadsheet incorporated into the package (e.g. EXCEL, MINITAB or SPSS). This approach demands less competence in computer usage, but can cause problems when transferring the data to another application or package. The construction of a rectangular matrix can become extremely laborious in the case where one is dealing with a dataset comprising a large number of species, of which many are scarce. Each species becomes a separate column in the data matrix, but generally most of the values will be zero. Mathematicians refer to matrices dominated by zeros as being sparse, and it is a characteristic of ecological data that they generate sparse matrices (since most species in typical ecosystems occur at low densities).

Table 1.5 gives a small example of ecological community data in two formats; firstly the list of species as recorded in a field notebook, and secondly the corresponding data matrix.

The author has faced editing a matrix of 60 species by 400 observations, in which most of the 24 000 cells were zero. Assembling and checking an array of this size from field notebooks is a huge task, and one better suited to computers than humans. There are programs available that store species data and export it in alternative formats, but as yet no package has gained acceptance as a standard solution. What has been developed is an alternative data format for ecological data that avoids the need for sparse matrices entirely. This is known as Cornell Condensed Format, and is described more fully in Chapter 7 because of its intimate links to the DECORANA package. (Cornell Condensed Format is also used by TWINSPAN and several multivariate ecological packages.) It is an efficient format for holding the sparse matrices typical of ecological data, but is, sadly, a difficult format for humans to read or verify.

One theme of this book is to examine underlying concepts, which are by their nature software-independent. Rather quaintly, Chatfield and Collins (1980) discuss the merits of punched cards versus magnetic media for data entry, while most of the current generation of students have never even encountered a punched data card. Precisely how you choose to assemble a data matrix depends on the software available to you, and on your personal preferences. However you choose to do this, there are some common points to be considered.

Column organisation

An important start is to distinguish between columns of measured data and columns that serve to identify the origin of the data. There will inevitably be some 'classification' variables, such as Site, Replicate, Date, Patient number, Depth, etc. which are vital for identifying where data came from but which must be excluded from actual analyses. It is good practice to have such classification variables as the first columns in a data matrix. It may be necessary to recode these into a format suitable for the package you are using – an example is given below in Tables 1.6 and 1.7. It should be obvious that classification variables must be excluded from multivariate analysis, although they may be used later as classification levels in ANOVA or similar univariate tests.

Jeffers (1995) refers to data which describe variables as metadata, and makes the point that these are essential to a proper analysis. He extends the definition of metadata to information about the sampling regime, including the identity of the sampler (where several people were involved), explanations for why missing values are missing, etc. and states that such metadata must be kept available to assist later analysis. In practical terms this describes good scientific practice of keeping a meticulous notebook recording precisely all details by which a set of data were acquired, and ensuring that such notebooks are preserved safely in perpetuity. If the data are held in a spreadsheet such as EXCEL, metadata can sensibly be held as a separate ply (sheet) held alongside the main data.

Missing values

An important point to consider is whether any of the measurements consist of missing values. Most packages will allow undefined data points to be held (usually portrayed as a period or '.'), but will be unable to include these points in most multivariate techniques. (It is possible to include missing data in multiple linear regression, although not necessarily very wise to do so.) There are two possible approaches to the handling of data containing missing observations. One can either exclude the variable(s) or the observation(s) containing missing data. This decision should be taken before undertaking the analysis, since

Table 1.6 An illustration of a common error in data format. This dataset gives data on plant growth in three different acid rain treatments, each treatment being replicated three times. This format is commonly used for writing data in notebooks or on blackboards due to its efficient use of space, but is unsuitable for data analysis by many commonly used statistical packages. This is because it violates the general rule that one column of data should correspond to one variable. In this case, one measured variable (plant biomass) is entered in three adjacent columns. It would be simple to enter data in this format into most standard packages by giving different names to each column, but any multivariate analyses that emerged would be invalid.

	pH		
	5.6	4.0	2.0
Radish biomass (g):	24.9	24.2	19.8
	22.7	28.3	18.7
	27.3	21.0	21.3

The correct general format for entering these into standard multivariate packages is shown below. Note that there are now three variables, with pH and replicate (REP) being explicitly defined for each observation, while the original three columns of data have been merged into one variable called MASS.

pH	REP	MASS
5.6	1	24.9
5.6	2	22.7
5.6	3	27.3
4.0	1	24.2
4.0	2	28.3
4.0	3	21.0
2.0	1	19.8
2.0	2	18.7
2.0	3	21.3

Table 1.7 The data from Table 1.6 recoded so that pH can be used as a classification variable, by addition of a new variable NEWPH, which takes values of 1, 2 or 3. On many packages it will be easier to analyse the data using NEWPH as the classification variable.

pH	NEWPH	REP	MASS
5.6	1	1	24.9
5.6	1	2	22.7
5.6	1	3	27.3
4.0	2	1	24.2
4.0	2	2	28.3
4.0	2	3	21.0
2.0	3	1	19.8
2.0	3	2	18.7
2.0	3	3	21.3

statistical packages may either try to carry on (by estimating the missing data), exclude the affected observations, or simply terminate the analysis.

Format errors

It is relatively common for data to be assembled in a format which makes sense to the researcher, but which subsequently proves to be incompatible with the requirements of the software packages used in its analysis. Most researchers encounter a problem of this nature early in their career, waste some time rearranging their data, and learn thereafter to save time by getting the formatting correct when data are first being assembled.

At the time of writing, most packages require data to be assembled in a matrix format, in which there is a 1:1 relationship between columns and variables. This means that one column contains data (numeric or other) about one variable, and that all the measurements on that variable are to be found in the same column. If this condition is not met it will generally prove to be impossible to perform a correct analysis on data (although generally no warning will be issued by the package).

A common example of such a formatting error can arise from copying a dataset directly out of a notebook or off a blackboard into a computer. It is quite common and normal when writing data down by hand to put the same variable in different columns. Thus when compiling data from a laboratory practical on (for example) the pH of simulated rain on growth of seedlings, one might record seedling growth under three columns as shown in Table 1.6.

It is easy to compile a data matrix this way, and indeed the widely used spreadsheet EXCEL actually requires this multicolumn format in order to perform analysis of variance on a variable. Unfortunately this format is not suitable for any of the standard multivariate packages because one variable is being defined in multiple columns. To continue the plant growth/acid rain analogy mentioned above, if biomass at pH 5.6 were to be presented in a different column to growth at pH 4, these two columns would then contain different variables as far as any package is concerned. They do not – they contain the same variable, in this case MASS. Note also that one row of data now comprises multiple observations – it is generally better if each observation is on a separate and distinct row. The correct format is shown in Table 1.6. Note that in addition to the dependent variable MASS, there are two columns of metadata that identify the observation uniquely (pH and REPLICATE).

Recoding data

As a software-specific caution, note that the pH treatments in the example shown in Table 1.6 take three values; 5.6, 4.0 and 2.0. Some packages will refuse to use this sort of irregular range if analysis of variance or related techniques are being considered, and depending on the software it might be essential to use alphanumeric characters or a simple 1–2–3 range. Since computers have huge storage capacity, and humans regularly become muddled with datasets, it is best if the original values of pH are retained after recoding. To continue the example above, the original pH/growth dataset could have been recoded so that pH 5.6 is replaced by 1, pH 4 by 2 and pH 2 by 3. This has been done by adding a new variable (NEWPH) which could be used as a classification variable in ANOVA, while retaining the original pH value to remind the human operator what the data actually mean. This insures against the weaknesses both of computers and of humans! Such recoding is often needed, but should only be done inside the statistical package once the data matrix has been assembled. How you undertake recoding is, of course, software specific. Table 1.7 shows the same dataset after recoding, with the new variable NEWPH.

Another important aspect of recoding concerns the inclusion of nominal data. Few packages allow non-numeric data to be included in any type of numerical analysis, although in some cases the information content could be valuable. For example, when dealing with taxonomic measurements on flowers, the flower colour is likely to be useful but problematic to include. The solution is to recode nominal data into a numeric scale. Chatfield and Collins (1980, p. 35) suggest that under these circumstances the coding is entirely arbitrary, so that (for example) three colours may be coded as one variable taking the values 1, 2 and 3. This is in fact invalid, since any analysis performed by computer will treat the colour coded as 3 as being three times the weight of the colour coded as 1. A better solution would be to introduce a new variable for each possible state, so that (continuing the three-colour example) colour would be defined by three new variables, BLACK, BROWN and RED, each taking the value 1 (if true) or 0. (Variables such as this which are either 0 or 1, meaning true or false, are often known as logical or Boolean variables, the latter name coming from George Boole who codified the principles of symbolic logic.)

Although this coding is correct, it should be noted that there is redundancy in the information. If it is known that colour is coded as one of three states, black, brown or red, it is sufficient to use two colour variables: BLACK ($=1$ when the colour was black, otherwise 0) and BROWN ($=1$ when the colour was brown, otherwise 0). The colour red is coded by BLACK $= 0$ and BROWN $= 0$.

On the face of it this redundancy seems trivial, but in fact a deep statistical problem could arise if three colour variables were created and all used in an analysis, since there would be co-correlation between them (due to the existence of redundant information). Data where two or more variables are highly correlated are known as **collinear** (or can be said to exhibit **collinearity**). Some multivariate analyses explicitly assume that variables are independent, and can give unreliable or misleading outputs if this condition is violated. The techniques that have problems with collinearity include multiple linear regression (Chapter 3), canonical correspondence analysis (Chapter 9), canonical correlation analysis and redundancy analysis (both Chapter 10). In these cases, collinearity must be removed by the removal of one (or more) variables from the analysis. The correct solution is to include all three colour variables in your data matrix, but only to incorporate two of them into the relevant analyses.

As an example, ter Braak (1986) explored the influence of three management techniques (addition of manure or a chemical fertiliser, or an unfertilised control) on a plant community using a multivariate technique called canonical correspondence analysis (Chapter 9). He represented the three soil management treatments by two new variables (MANURE and FERTILISER), each taking the value 1 or 0. The unfertilised control situation was defined by MANURE $= 0$ and FERTILISER $= 0$. As a general principle, a nominal variable with N states can be defined by $N - 1$ variables, each taking the value of 0 or 1.

1.5.4 Preliminary inspection of the data

It may seem odd to suggest inspecting data by eye before invoking the powerful analytical techniques described in this book. In fact the prudent researcher will spend at least as long verifying a data matrix as analysing it. This comes from the piece of computer science folklore known by its acronym GIGO – Garbage In Garbage Out. It is a common experience for a researcher to run a multivariate analysis and generate a plausible-looking output which leads to a sensible interpretation of the data. The researcher then realises that there is a serious problem with the matrix or the analysis, corrects this and generates a totally different but equally sensible looking output. The first output was in fact spurious rubbish, but its fraudulence was not immediately obvious. (One example among many comes when missing data receive special coding. Chatfield and Collins (1980) give the example of a dataset profiling car accidents, in which many women drivers declined to give their age.

The missing ages were entered as 99, resulting in an analysis which showed a surprisingly high frequency of accidents involving cars driven by very old women!)

The problem is that multivariate techniques are excellent tools for detecting patterns within data, and if one of the patterns happens to concern an odd, incorrectly entered data point then this will be picked up. Figure 1.3 shows graphically the dangers of allowing an erroneous data point to enter a multivariate analysis. It shows the summary of a multivariate analysis of a rather patternless dataset. (The actual analytical technique used is principal components analysis, described in Chapter 6, although most other multivariate techniques are equally sensitive.) The important point here is to realise that this technique searches for useful descriptions of the shape of a dataset, and presents this in a figure known as an ordination diagram (Chapter 4). The same dataset was reanalysed after changing just one number in one observation by a factor of ten – in other words misentering its decimal place. The ordination diagram of this new dataset is so different as to be unrecognisable. The analysis successfully detected a pattern – namely that one data point had an unexpected value. The difference between the two ordination diagrams in Figure 1.3 is caused by one slip of one decimal place in one observation, and should serve as a warning to check all data carefully before undertaking multivariate analysis.

Multivariate techniques are always sensitive to outliers, and even a correctly entered observation may skew an analysis if it happens to have an extreme value. It is highly recommended that data be inspected for outliers before undertaking any multivariate analyses.

Most univariate statistical techniques assume that data come from a population approximating to the **normal distribution** (also known as the Gaussian distribution), and there are standard tests available to assess how closely data fit to this assumption. Similarly, many multivariate techniques assume that all variables entered come from the multivariate version of the normal distribution, unsurprisingly known as the **multivariate normal**

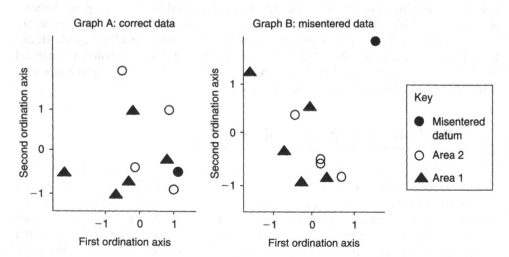

Figure 1.3 An example of the sensitivity of multivariate analyses to errors in data entry. Here the same dataset, comparing soil properties from two areas, has been subjected to the same analysis. (The analysis used was the most commonly used ordination technique, principal components analysis.) The only difference is that in Part B, one data point has been misentered by a mistyped decimal point. This changes the entire analysis – the two diagrams are unrecognisably different. The removal of a single outlier can make an equally dramatic difference to a multivariate analysis.

distribution. Unfortunately, there are no standard tests to assess the fit of a dataset to this distribution, and the simplest solution is to check whether each variable in a multivariate set is itself normally distributed. Consequently the prudent researcher will check the distribution of each variable carefully, and consider whether to apply a transformation (commonly the logarithmic transformation $X' = \log(x)$) to control the data distributions. For population data, the simple logarithmic transformation is inapplicable since zero values will occur, and the logarithm of zero may not be calculated. The preferred transformation in such cases is to use $X' = \log(X + 1)$.

The process of deciding whether data approximate to a normal distribution can be approached in several ways. This is demonstrated using data on biomass harvested from a fertiliser trial (Table 1.8). A simple start is to inspect a cumulative p-p plot, which is a graph that shows the cumulative values of the numbers (starting from the smallest) along with the cumulative shape that would be expected if the data were indeed derived from a normal distribution (Figure 1.4). Visible deviations from the line are a warning that the data may not approximate normality. These can be followed up by a formal test which returns a probability of the data being normal, such as the Kolgomorov–Smirnov one-sample test. Figure 1.4 shows that both the p-p plot and the Kolgomorov–Smirnov statistic indicate that the biomass data in Table 1.8 deviates significantly from normality. Figure 1.5 repeats this exercise after logarithmic transformation (in the case of the data in Table 1.8 by logarithms base 10, but any other logarithmic base would do equally well) and shows that the transformed data approximate much better to the normal distribution. Having established this it would be prudent to continue analyses using the log-transformed values (when it can be shown that the data exhibit a strong response to the fertilisers, especially the organic fertiliser).

Once the user is confident about the integrity and distribution of the data, there are further useful preliminary approaches to data inspection. This should include visual examination of trends. Using computers, graphs can be plotted and overlain with symbols in seconds. A good subjective assessment of trends within data can be obtained by looking at graphs describing how the commoner species/major elements behave, although confirming

Table 1.8 Data to illustrate the process of normality testing. These concern the dry mass of vegetation harvested from an experiment in which three fertiliser treatments were applied to the industrial waste FGD gypsum (unpublished data). This would be a classic example of a situation requiring analysis of variance (ANOVA), but the data prove to follow a skewed distribution that is ameliorated by log transformation (Figures 1.4 and 1.5).

Fertiliser	mass harvested, g	$\log_{10}$(mass)	Fertiliser	mass harvested, g	$\log_{10}$(mass)	Fertiliser	mass harvested, g	$\log_{10}$(mass)
None	0.23	−0.64	Chemical	3.45	0.54	Organic	5.68	0.75
None	0.28	−0.55	Chemical	3.88	0.59	Organic	7.03	0.85
None	0.38	−0.42	Chemical	4.82	0.68	Organic	9.71	0.99
None	0.54	−0.27	Chemical	5.78	0.76	Organic	10.05	1
None	0.68	−0.17	Chemical	8.16	0.91	Organic	10.51	1.02
None	0.79	−0.1	Chemical	8.5	0.93	Organic	11.02	1.04
None	1.25	0.1	Chemical	9.34	0.97	Organic	11.56	1.06
None	1.51	0.18	Chemical	10.17	1.01	Organic	13.63	1.13
None	1.82	0.26	Chemical	10.28	1.01	Organic	29.26	1.47
None	1.95	0.29	Chemical	13.15	1.12	Organic	44.4	1.65
None	2.13	0.33	Chemical	14.78	1.17	Organic	51.18	1.71
None	2.39	0.38	Chemical	16.16	1.21	Organic	51.8	1.71
None	2.77	0.44	Chemical	26.93	1.43	Organic	65.16	1.81
None	3.65	0.56	Chemical	30.01	1.48	Organic	71.27	1.85
None	5.13	0.71	Chemical	66.2	1.82	Organic	87.13	1.94

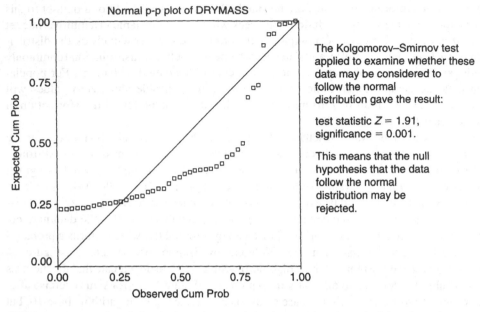

The Kolgomorov–Smirnov test applied to examine whether these data may be considered to follow the normal distribution gave the result:

test statistic $Z = 1.91$, significance $= 0.001$.

This means that the null hypothesis that the data follow the normal distribution may be rejected.

Figure 1.4 A p-p plot of the dry mass harvest data, showing a serious deviation from the normal distribution. Ideally normal data will lie on the black diagonal line running from bottom left to top right.

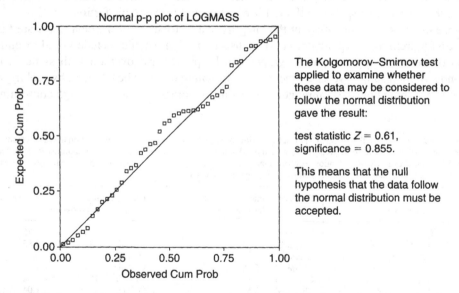

The Kolgomorov–Smirnov test applied to examine whether these data may be considered to follow the normal distribution gave the result:

test statistic $Z = 0.61$, significance $= 0.855$.

This means that the null hypothesis that the data follow the normal distribution must be accepted.

Figure 1.5 A p-p plot of the dry mass harvest data after logarithmic transformation, showing that the transformed data fit much better to the shape of the normal distribution.

the validity of these impressions will probably involve some of the powerful multivariate techniques described in this book.

Finally, there is good reason to construct and examine the correlation matrix of a data set. This is explained in Chapter 6, and simply involves calculating the correlation coefficient between each pair of variables in turn. (This matrix, and a corresponding matrix of probabilities, are routinely supplied by any standard statistical package.) The reason for doing this is to check whether some of the variables are closely correlated, as is frequently the case in environmental data. As explained above, collinear data cause problems for some multivariate techniques (MLR, CCA, COR and RDA). If these techniques are to be applied to collinear data, the solution is to remove one or more variables from the dataset, until the remaining variables are uncorrelated. By contrast, indirect ordination techniques (Chapters 5–7) may safely be used to simplify collinear data, indeed are ideally suited for so doing.

1.6 The sample datasets

Throughout this book we will return to reanalysis of several environmental datasets, in order that the reader may see how different analyses applied to the same dataset can give differing but generally complementary insights into the patterns concealed within the data.

1.6.1 The Liphook pine forest mycorrhizal fungal succession

This dataset (Table 1.9) consists of counts of the total number of toadstools within $40\,m^2$ plots for each of five years (1986–1990) produced by mycorrhizal fungi in plots of Scots Pines growing in a heathy soil in the south of England. The pines were being subjected to accurately controlled levels of pollution by the gases sulphur dioxide (SO_2) and ozone (O_3) to determine the effects of chronic air pollution on forest health (Shaw *et al.*, 1992). The mycorrhizal fungi were being studied because they play a crucial role in forest nutrient cycling: the fungi ensheath the roots of trees and supply both protection and nutrients, in return for sugars supplied by the tree. It was anticipated that pollutant injury to the forest

Table 1.9 The Liphook pine forest toadstool data 1986–1990. Abbreviations: Bf – *Boletus ferrugineus*, Cs – *Cortinarius semisanguineus*, Gr – *Gomphidius roseus*, Il – *Inocybe lacera*, Lp – *Laccaria proxima*, Pi – *Paxillus involutus*, Sb – *Suillus bovinus*, Sl – *Suillus luteus*, Sv – *Suillus variegatus*. The shaded cells pick out two observations (from plot 1 in 1989 and 1990) which will be shown by several ordinations to be outliers, with a very different community to the rest of the dataset.

Obs	Year	pl	Bs	Cs	Gr	Il	Lp	Lr	Pi	Sb	Sl	Sv
1	86	1	0	0	0	0	280	0	1	0	0	0
2	86	2	0	0	0	0	171	0	51	0	0	0
3	86	3	0	0	0	0	12	0	29	0	0	0
4	86	4	0	0	0	0	40	0	2	0	0	0
5	86	5	0	0	0	0	2	0	9	0	0	0
6	86	6	0	0	0	0	34	0	7	0	0	0
7	86	7	0	0	0	0	28	0	18	0	0	0
8	87	1	0	0	1	0	1021	0	1	5	0	2
9	87	2	2	0	21	0	547	0	27	42	0	0
10	87	3	0	0	11	0	151	0	31	31	0	0
11	87	4	4	0	64	0	438	0	6	119	0	0
12	87	5	0	0	0	0	99	0	34	0	0	0
13	87	6	1	0	0	0	261	0	4	0	0	0
14	87	7	0	0	2	0	242	0	28	22	0	0
15	88	1	0	11	7	15	1265	0	0	299	3	102
16	88	2	3	0	30	0	440	0	3	350	1	15
17	88	3	2	0	63	1	229	0	0	394	0	0
18	88	4	9	1	227	0	883	0	0	1044	7	18
19	88	5	0	0	0	0	230	0	13	34	0	0
20	88	6	2	0	3	0	380	0	1	128	0	6
21	88	7	2	0	100	0	441	0	3	733	0	10
22	89	1	0	44	8	1	19	0	0	568	10	295
23	89	2	2	0	59	0	26	0	2	595	0	97
24	89	3	3	0	196	0	9	0	2	1255	1	3
25	89	4	10	0	427	0	59	0	1	1434	1	24
26	89	5	0	0	11	0	38	2	14	331	0	5
27	89	6	0	0	20	0	24	0	2	313	0	76
28	89	7	2	1	175	0	7	0	0	1327	0	34
29	90	1	1	111	4	6	24	24	2	36	14	116
30	90	2	1	4	11	0	24	4	2	215	0	9
31	90	3	0	7	70	0	49	12	3	655	2	0
32	90	4	3	17	75	0	66	1	1	538	7	2
33	90	5	0	0	13	0	44	5	19	159	0	5
34	90	6	2	0	6	0	42	0	11	180	0	22
35	90	7	0	12	43	0	17	1	2	644	0	0

ecosystem might manifest itself as reductions in the production of mycorrhizal toadstools, hence the desire to collect these data.

The pollutant gases were released in five out of seven experimental plots: the treatments are defined in Table 1.10. For more technical background see Shaw (2000). The data have been condensed down to one total per species per treatment plot per year. In fact the dominant pattern within the data is that of an ecological succession, with the species changing as the young trees grew (Figure 1.6). The early years were dominated by the Deceiver *Laccaria proxima* and the common roll rim *Paxillus involutus*, while later *Suillus bovinus* became the commonest species. Superimposed on this are differences between the experimental plots, although they do not seem to be related to the fumigation treatments and instead probably represent soil differences.

Table 1.10 The fumigation treatments used in the Liphook experiment, and their corresponding annual mean concentrations (in parts per billion) for the period 1988–1990.

Plot number	SO$_2$ treatment	O$_3$ treatment
1	ambient, 4 ppb	ambient, 25 ppb
2	high, 22 ppb	ambient, 25 ppb
3	low, 13 ppb	high, 30 ppb
4	ambient, 4 ppb	high, 30 ppb
5	high, 22 ppb	high, 30 ppb
6	low, 13 ppb	ambient, 25 ppb
7	ambient, 12 ppb	ambient, 25 ppb

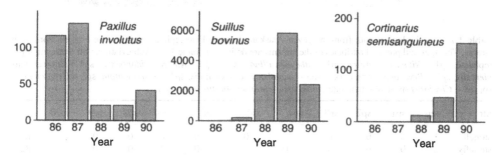

Figure 1.6 The Liphook pine forest fungal dataset is dominated by a successional change from a species-poor community dominated by *Paxillus involutus* to one dominated by species of *Cortinarius* and *Suillus*.

1.6.2 Life in Alaskan streams: chemical conditions and freshwater crustacea

These data come from South eastern Alaska, where glaciers dominate the landscape, but have long been in retreat. Where glaciers melt, streams develop, and in due course these are colonised by life. The slow retreat of the glaciers has been well studied, so it is possible to estimate the age of stream sections with a high level of confidence. The subjects of this study are the small crustaceans – the microcrustacea – which prove remarkably able to colonise new waterways against the flow of newly formed streams. Data presented here come from seven sites, ranging in age from 18 to approximately 1400 years.

The seven streams were sampled by a standard kick-sampling technique, in each case five replicate samples being taken. Physico-chemical data were also collected, but only one sample was analysed per stream.

The data are presented in Tables 1.11 and 1.12. Preliminary inspection of the ecological data show very different communities between the youngest and oldest streams, with a species-poor community dominated by *Maraenobiotus insignipes* (a Harpacticoid copepod) in the youngest stream (stonefly, 18 years), a more diverse community in the middle-aged sites (50–150 years), and a distinct assemblage in the oldest site (Carolus, estimated at 1377 years). This pattern is shown graphically in Figure 1.7 for a subset of species.

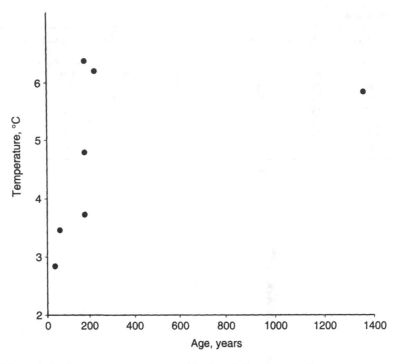

Figure 1.8 Variation in water temperature within the Alaskan streams dataset.

Table 1.13 Correlations between the environmental variables measured in the Alaskan streams dataset (using Pearson's correlation coefficient with 5 df). A significant value ($p < 0.01$) is shown in bold.

	Age, years	Pfankuch index	%cwd	Temp, C	turbidity	conductivity	alkalinity
Age, years	1.00						
Pfankuch index	0.60	1.00					
%cwd	0.27	0.09	1.00				
Temp, C	0.41	0.48	0.59	1.00			
Turbidity	−0.34	−0.31	−0.37	−0.60	1.00		
Conductivity	0.03	−0.44	0.62	0.24	0.06	1.00	
Alkalinity	0.01	−0.37	0.74	0.40	−0.09	**0.97**	1.00

1.6.3 The Wimbledon Common spoil mound

These data come from a site on Wimbledon Common (London), where sandy gravelly soil laid down in the Quaternary has allowed the formation of a lowland heath. This has developed typical heathland flora (acid-loving grasses with heather) and soil (low pH and dissolved salts, high in organic matter). On top of this were dumped many thousands of tonnes of London clay, excavated when the adjacent A3 was widened in the 1960s. (It seems remarkable that permission for this dumping was ever obtained.) This clay is slightly alkaline, retains rather higher levels of dissolved salts, and has developed a very different plant community to the heath.

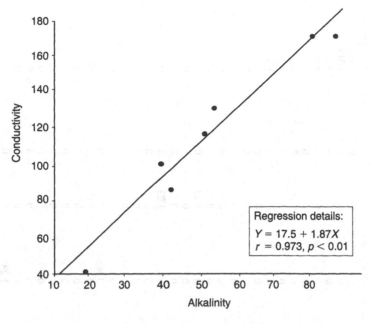

Figure 1.9 The linear relationship between conductivity and alkalinity in the Alaskan streams dataset.

Table 1.14 The data from Table 1.11 after transformation by log(x + 1).

	Nitocra hibernica	Atheyella illinoisensis	Atheyella idahoensis	Bryocamptus hiemalis	Bryocamptus zschokkei	Acanthocyclops vernalis	Alona guttata	Graptoleberis	Chydorus	Macrothricidae	Maraenobiotus insignipes
stonefly	0	0	0	0	0	0	0	0	0	0	1.813
stonefly	0	0	0	0	0	0	0	0	0	0	1.519
wolf pt	0	0	0	0	0	2.41	0	0	0	0	0
wolf pt	0	0	0	0	0	2.585	0	0	0	0	0
wolf pt	0	0	0	0	0	2.848	0	0	0	0	0
wolf pt	0	0	0	0	0	3.558	0	0	0	0	0
wolf pt	0	0	0	0	0	1.908	0	0	0	0	1.23
tyndall	0	0	0	2.436	2.528	1.23	0	0	0	0	0
tyndall	0	0	0	2.053	2.585	0	0	0	0	0	0
tyndall	0	0	0	1.69	2.773	1.23	0	0	0	0	0
tyndall	0	0	0	1.813	2.286	0	0	0	0	0	0
tyndall	0	0	0	2.111	2.161	0	0	0	0	0	0
berg n	0	0	1.69	1.519	1.519	0	0	0	0	0	0
berg n	0	0	1.908	1.23	0	0	0	0	1.23	0	0
berg n	0	0	0	2.161	1.23	0	0	0	0	0	0
berg n	0	0	1.813	2.461	0	0	0	0	1.23	0	0
berg n	0	0	1.69	3.118	1.69	1.519	0	0	0	0	0
berg s	0	0	1.23	1.23	1.519	0	0	0	0	0	0
berg s	0	0	0	0	1.69	1.23	0	0	0	0	0
berg s	0	0	0	1.23	1.69	1.23	0	0	0	0	0
berg s	0	0	0	0	1.519	1.23	0	0	0	0	0
berg s	0	0	0	1.519	1.23	0	0	0	0	0	0
rush pt	0	0	0	0	0	1.23	0	0	0	0	0
rush pt	0	0	0	1.519	1.519	0	0	0	0	0	0
rush pt	0	0	0	1.23	0	1.69	0	0	0	0	0
rush pt	0	0	0	1.23	1.519	0	0	0	0	0	0
carolus	1.908	0	0	0	0	0	1.23	0	0	0	0
carolus	2.207	1.69	0	0	0	1.519	0	0	0	0	0
carolus	2.567	1.519	0	0	0	1.813	0	1.23	0	0	0
carolus	1.519	1.23	0	0	0	1.23	1.519	0	0	1.23	0
carolus	1.69	0	0	0	0	1.813	0	1.23	0	0	0

This site has provided an examplar dataset in which two essentially different sets of organisms occupy two different environments, although living side by side. This is a useful 'test bed' to explore multivariate analyses, since the answers that emerge can easily be related back to two sharply contrasting systems. These data have previously been introduced as Table 1.4.

2

Measurements of ecological diversity

2.1 The concept of diversity

Diversity is a deceptive concept, appearing on the surface to be simple, but revealing depth and complexity on closer thought. Magurran (1991) likens the concept of diversity to an optical illusion; the more it is examined, the less clear it becomes. In daily speech one tends to talk about diversity as meaning the number of types or categories in a sample. Thus one might state that an old fashioned town centre with many types of shop is more diverse than modern shopping areas containing a few types of very large supermarkets. Ecologists have long been interested in measuring the diversity of ecosystems, going back to the assertion by McArthur (1955) that diversity within an ecosystem is a predictor of its stability and the efficiency of its ecosystem functions. (This idea, known as the diversity–stability hypothesis, came in for severe criticism by Goodman (1975), and has remained a source of contention since – e.g. Tilman *et al.*, 1996, Chapin *et al.*, 1998.)

The distribution of diversity on a spatial scale is described within ecology as comprising alpha, beta and gamma diversity. These measurements relate to the way that diversity varies with sampling scale. Alpha diversity is the diversity of organisms within a selected habitat or sample, and the indices we will cover in this chapter are formally indices of alpha diversity. Beta diversity is an index of the rate of increase of alpha diversity as new habitats are sampled, so is a measure of the turnover of species along a spatial gradient. We will not explore beta diversity further, except to note that one of the properties of the ordination package DECORANA (Chapter 7) is to rescale data so that the rate of turnover along its axes is approximately constant – which is equivalent to producing a constant beta diversity along its ordination axes. Finally, gamma diversity is the full diversity (species richness) of an entire sampled landscape or gradient. It will be noted that these definitions of alpha, beta and gamma diversity are scale dependent, so that a patch size that a mammal ecologist would consider to be one habitat (measuring alpha diversity) would be a mosaic of microhabitats which a microbial ecologist might count as containing gamma diversity.

We can now ask what properties of the community may be measured to indicate its alpha diversity. The first and most obvious measurement to make is the count of the total number of species within the sample. This is known as the species richness, and is a valid index of diversity in its own right. However a simple total count of species gives no information about their relative frequencies, and other indices of diversity are constructed which include a measure of the evenness with which species are distributed. Consider two ecosystems, in

Habitat 1 (low diversity)   Habitat 2 (high diversity)

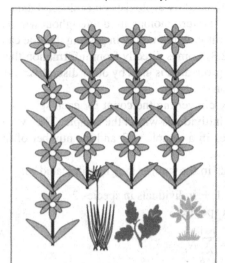 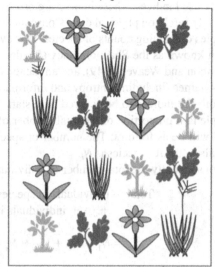

Figure 2.1 The difference between species richness and diversity. The diagram shows two samples, both containing the same number of species and individuals (16 individuals from four species), but with a clearly different balance of species composition.

both of which a systematic survey turned up four species (Figure 2.1). These two sites would have equal richness, but the patterns of occurrence could be very different. At one extreme each species might occur in exactly equal numbers – this is a high-diversity situation. At the other extreme one species might completely dominate, so that the other four were only found on one occasion each during the whole sampling programme. This still gives it a richness of four species, but to a good approximation there is only one species in that ecosystem – hence it is a low-diversity situation.

The aim of diversity indices is to encapsulate these concepts in one number, which may be used to compare the relative diversity of two or more situations. Unlike the other techniques described in this book, diversity indices are relatively quick and simple to calculate using a scientific pocket calculator or a spreadsheet. In fact standard statistical packages do not at present routinely supply diversity indices, although these can be found on a number of packages aimed specifically at ecologists (Appendix 3).

Thus the concept of diversity contains two elements: richness and balance. The relative importance of these two elements is not defined, and there are potentially an infinite number of different mathematical functions that could plausibly be used as diversity indices by encapsulating different aspects of the balance between these two elements. By way of example, Hill (1973a) gives one family of functions which could generate infinitely many different viable diversity indices.

The study of approaches to measurements of diversity is consequently a large one, and it is not my intention to provide a comprehensive coverage in this chapter. The serious reader is referred to Magurran (1991) for a greatly more detailed coverage of the entire subject, including estimation of total species richness (including undiscovered species) and fitting mathematical functions to species distributions.

The aim of this chapter is more modest: to introduce the reader to a few of the most commonly used indices. These indices are primarily used by ecologists on ecological datasets, and only rarely on physical data. Consequently the terminology used throughout will be of samples containing counts of species. However it is worth noting in passing that the calculation known as the Shannon index (Section 2.3) comes directly from information theory (Shannon and Weaver, 1949), and an identical procedure is used by other disciplines under other names (including entropy and information content).

Both the indices to be covered here start by converting the count for each species in a sample to a proportion of the total number of individuals within the sample. The symbols used will be as follows: The number of species in a sample is S, and the number of individuals in the ith species is N_i.

Consequently the total number of individuals in the sample may be calculated as:

$$\text{Total} = \text{individuals in species } 1 + \text{individuals in species } 2$$
$$+ \cdots + \text{individuals in species } S$$

$$= N_1 + N_2 + \cdots + N_s = \sum_{i=S}^{i=1} N_i$$

which may be written as $\sum N$ for convenience.

The proportion made up by species i (denoted by p_i) is given by:

$$p_i = \frac{N_i}{\sum N} \tag{2.1}$$

For most of the diversity indices considered below (the exception being Brillouin's index), the actual numbers of individual organisms in any sample could all be multiplied by 2, 10 or any other non-zero scaling factor without altering the indices generated – what matters is the relative balance between each species. It should be noted that these definitions imply that it must be meaningful to add data together. Diversity indices are generally applied to species counts, although other continuous data such as percentage cover could also be used. Diversity indices are not applicable for nominal data (flower patterns defined by sets of colours, or areas classified by land use types), nor are they applicable to ordinal data, such as plant communities scored on a DAFOR scale (Section 1.5).

Of the many different approaches to quantifying diversity, only the two most commonly and easily used will be described in depth here, namely the Simpson and Shannon indices. The mathematical formulation of a further selection will be given without detailed explanation at the end of this chapter.

2.2 The Simpson index

This index dates from the work of Simpson (1949), and has been widely used due to its conceptual simplicity and ease of calculation. This index relates to the probability that two consecutive random samples from a population will find the same species. Clearly if a community consists of just one species, the probability that two consecutive samples will find the same species must be 1.0. Equivalently, in a highly diverse community this probability will be approximately zero.

The probability that a random sample from a population will pick out a given species can be assumed to be equal to that species' contribution to the whole population. (This index ignores any logistical problems in sampling, and treats all individuals in all species as being identical balls hidden in a bag.) Hence the probability of sampling species i is simply that species' proportion of the total, given by equation 2.1:

$$p_i = \frac{N_i}{\sum N}$$

Assuming that the sampling programme does not disrupt the population, the probability of sampling species i in two consecutive samples is found as follows:

$$p(\text{sampling species } i \text{ twice}) = p_i \times p_i \qquad (2.2)$$

Note that technically equation 2.2 is only valid when an individual is 'sampled' then returned to the population, so is available for resampling. A more realistic model is one in which individuals are removed from the population by sampling, in which case equation 2.2 should be rewritten as:

$$p(\text{sampling species } i \text{ twice}) = \frac{N_i(N_i - 1)}{\sum N(\sum N - 1)} \qquad (2.3)$$

In practice this modification makes little difference, and is rarely used.

The probability of sampling any species twice in two consecutive samples can be found as the sum of probabilities for each individual species, as follows:

$$p(\text{sampling any species twice}) = p(\text{sampling species 1 twice}) + p(\text{sampling species 2}$$
$$\text{twice}) + \cdots + p(\text{sampling species } N \text{ twice})$$

$$= \sum_{i=1}^{i=S} (p_i \times p_i) \qquad (2.4)$$

and is the basis of Simpson's index. Clearly if there is only one species, $p_i = 1$ hence $\sum(p_i \times p_i) = 1$. This is the zero diversity condition. As the number of species tends to infinity, $\sum(p_i \times p_i)$ tends to zero, which is the high-diversity condition.

The difficulty with this basic version of the Simpson's index is now apparent: zero diversity scores 1 while infinite diversity scores 0. Although this is mathematically acceptable it is counter-intuitive, and Simpson's index is usually altered to reverse this arrangement.

The standard expression of Simpson's index is to calculate the probability that two consecutive samples will be of different species, which is $1 -$ (the probability that the samples will be of the same species), so (using equation 2.4) is calculated as follows:

$$D = 1 - \sum(p_i \times p_i) \qquad (2.5)$$

where D is the standard symbol for Simpson's index.

Being a probability, this version of Simpson's index ranges from zero (meaning zero diversity, i.e. the sample only contains a single species) to unity (meaning infinite diversity).

An alternative (less satisfactory) solution is to calculate the reciprocal of the original diversity index:

$$D = \frac{1}{\sum(p_i \times p_i)}$$

This index has a lower limit of 1.0 (corresponding to zero diversity), but has no upper limit.

2.3 The Shannon index

This is probably the most commonly used diversity index, although its interpretation is less simple than the Simpson index. It is certainly the most commonly misnamed! A frequent but erroneous name applied to this index is the Shannon–Weaver index (Krebs, 1985). An alternative, acceptable name for this index is the Shannon–Wiener index. This index, conventionally given the symbol H, arises from studies of information theory undertaken by Claude Shannon in the Bell Telephone Laboratories in the 1940s (Shannon and Weaver, 1949) although it was introduced into ecology by Margalef (1958).

The basis of the index relates to calculations of the maximum possible information content of data, and is calculated as:

$$H' = -\sum p_i \times \log(p_i) \tag{2.6}$$

where H' is the standard symbol for the Shannon index, and $\log(p_i)$ is the logarithm of p_i, the proportion of species i. Since this proportion will always be less than 1.0, the logarithm will always be negative (except for the limiting case where the community consists of just one species, where $p_i = 1.0$ and its logarithm is 0.0). The minus sign at the start of equation 2.6 is introduced to ensure that the calculated diversity is always a positive value.

For the simplest case (a community with only one species, where by definition $p_i = 1.0$) the diversity is zero. For more complex communities, the index reflects both the number of species and their relative balance. For a community with S species, the maximum possible value of the Shannon index is $\log(S)$ – this occurs when all species occur at equal frequency.

One complication of the Shannon index is that different workers have used different bases for the logarithm. The original version used logarithm base 2, in which case the diversity index can be interpreted as an information content in binary digits (bits). This is rather arcane for ecological studies, and logarithms base 10 are commonly used. The main reason for this is most computer packages and pocket calculators do not give logarithms base 2, only base 10 and natural (Naperian) logarithms base e, where e = 2.71828 .... The conversion between logarithms of different bases is simple:

$$\text{Log}_a(X) = \frac{\text{Log}_b(X)}{\text{Log}_b(a)} \tag{2.7}$$

where $\text{Log}_a(X)$ means the logarithm base a of the number X.

Combining equations 2.6 and 2.7 gives:

$$H'_{(\text{base 2})} = \frac{-\sum p_i \times \log_{10}(p_i)}{\log_{10}(2)} = 3.3219 \times H'_{(\text{base 10})}$$

A good alternative, which avoids the problems of defining logarithmic base entirely and imposes an allowable range of 0–1, is to calculate the ratio of the calculated diversity with the maximum possible diversity for the number of species found. This is known as the equitability (E), and is given as:

$$E = \frac{H'}{H_{\max}} = \frac{-\sum p_i \times \log(p_i)}{\log(S)} \tag{2.8}$$

where E is the equitability and S is the number of species. This index is unchanged whatever sort of logarithm is used, and reflects the evenness of species distribution within the sample. An equitability near zero shows the community to be dominated by one species, while a value near 1.0 shows it to have an equal balance between all species.

2.4 Jack-knifing

Diversity indices do not come with estimates of variability: in theory one community of organisms has one value for any chosen diversity index. In practice a researcher will repeatedly resample the community, and accumulated observations from many visits permit the calculation of mean, standard deviation and other standard statistical tools. It is a serious error (known as pseudoreplication, see Section 1.5.2) to believe that consecutive observations on the same biological community are independent, but these estimates of variation may be of interest to the researcher.

Jack-knifing is an extension of the resampling process, performed by computer using the completed final dataset (Efron 1979). It is a computing-intensive tool which can be used to obtain estimates of the variability within parameter estimates in a wide range of settings, including diversity indices. The procedure is to repeatedly reanalyse the data, excluding one observation at a time. This way a dataset with N observations will generate N different estimates (known as pseudovalues) of the diversity. These values are then treated as a dataset in their own right, from which a mean and estimates of variance may be calculated.

2.5 Examples

2.5.1 Example: The Wimbledon Common dataset

We start by calculating diversity indices for the data listed in Table 1.4, comparing plant species diversity on the lowland heath of Wimbledon Common with an adjacent spoil mound of negligible conservation interest. Species richness is a simple but useful start to the analysis; the heathland only held three plant species, while the spoil held five.

The manual calculations needed to derive the Simpson and Shannon indices are given in Table 2.1 (taking as a worked example the first observation from Table 1.4). This starts by converting each observation into p_i, the proportion of the total of the recorded observations for its sample. It is a useful arithmetic check to verify that $\sum p_i = 1.0$ for each sample. The calculations for $\sum(p_i \times p_i)$ and $\sum p_i \times \log_{10}(p_i)$ are now easily made, giving a final value for Simpson's index of 0.42 and the Shannon index of 1.02 (base 2) or 0.31 (base 10). Final figures are listed in Table 2.2.

These figures show slightly higher species richness on the spoil mound but little evidence for differences in diversity between these habitats, on the basis of this very small dataset. Analysis of variance by habitat showed species richness to be significantly higher on the

Table 2.1 The calculations of diversity indices, using the first line of data from Table 1.4.

	Raw data	p_i	$p_i \times p_i$	$-p_i \times Log_{10}(p_i)$
Achillea millefolia	0	0.00	0.00	0.00
Arrhenatherium elatius	0	0.00	0.00	0.00
Calluna vulgaris	5	0.04	2e−3	0.06
Deschampsia flexuosa	25	0.23	0.05	0.15
Festuca rubra	80	0.73	0.53	0.10
Heracleum sphondylium	0	0.00	0.00	0.00
Trifolium repens	0	0.00	0.00	0.00
Vicia sativa	0	0.00	0.00	0.00
Total	110	1.0	0.58	0.31
Species richness	3			
Simpson's index	0.42 (calculated as 1−0.58)			
Shannon's index (base 10)	0.31			
Shannon's index (base 2)	1.02 (calculated as 0.31/log10(2))			
Equitability	0.34 (calculated as 0.31/log10(8), from 8 species)			

Table 2.2 Final diversity values for the data from Table 1.4.

Sample	Species richness	Simpson's D	Shannon's H (base 10)
Heath 1	3	0.42	0.31
Heath 2	2	0.17	0.14
Heath 3	3	0.61	0.44
Heath 4	3	0.49	0.33
Spoil 1	4	0.32	0.28
Spoil 2	3	0.27	0.23
Spoil 3	5	0.51	0.40
Spoil 4	4	0.53	0.37

spoil mound, but not the two derived diversity indices. In fact more thorough surveys of the site consistently find more species richness on the spoil mound site, despite the heathland being a far more valued habitat due to its antiquity. The explanation for this lies in the age, cultural connotations and specialised animal communities of lowland Calluna heath. This should act as a note of caution against over-reliance on objective diversity indices in deciding the conservation importance of a habitat (Margules and Usher, 1981; Margules, 1986).

2.5.2 Calculating diversity indices on a spreadsheet

Most researchers will have access to spreadsheets such as the EXCEL package, which can readily be used to calculate these diversity indices. Although it is necessary to specify each calculation in a separate cell on the spreadsheet, this is readily achieved by copying and pasting one cell for each operation. The procedure for calculating the Simpson index is illustrated in Figures 2.2 and 2.3, which show first the raw numeric spreadsheet, and (separately) the commands typed into each cell. The three cells circled should be copied and pasted into all the cells up to their grey boundaries. This starts with a list of species names (optional, but useful for data verification), followed by the frequencies of occurrence in each habitat. These are followed by a demarcation line of greyed cells, which separate the raw data from the cells of calculations. (This demarcation line makes no difference to the

Spreadsheet to calculate the Simpson diversity index

	raw data			p_i			p_i^2	
	heath 1	spoil 1		heath 1	spoil 1		heath 1	spoil 1
Achillea millefolia	0	10		0	0.116		0	0.0135
Arrhenatherium elatius	0	70		0	0.814		0	0.6625
Festuca rubra	80	0		0.727	0		0.5289	0
Calluna vulgaris	5	0		0.045	0		0.0021	0
Deschampsia flexuosa	25	0		0.227	0		0.0517	0
Heracleum sphondylium	0	0		0	0		0	0
Trifolium repens	0	5		0	0.058		0	0.0034
Vicia sativa	0	1		0	0.012		0	0.0001
sum:	110	86		1	1		0.5826	0.6796
				Simpson's diversity:			0.4174	0.3204

Figure 2.2 The spreadsheet showing the calculation of Simpson's index for two observations within the Wimbledon Common dataset. (This sheet shows the actual numeric values calculated.)

	A	B	C	D	E	F	G	H	I
1		raw data				p_i			p_i^2
2		heath 1	spoil 1		heath 1	spoil 1		heath 1	spoil 1
3	Achillea millefolia	0	10		=B3/B$12	=C3/C$12		=E3*E3	=F3*F3
4	Arrhenatherium elatius	0	70		=B4/B$12	=C4/C$12		=E4*E4	=F4*F4
5	Festuca rubra	80	0		etc	..		etc	..
6	Calluna vulgaris	5	0		..	..		..	..
7	Deschampsia flexuosa	25	0		..	..		..	..
8	Heracleum sphondylium	0	0		..	..		..	..
9	Trifolium repens	0	5		..	..		..	..
10	Vicia sativa	0	1		..	..		..	..
11									
12	sum:	=SUM(B3:B10)	=SUM (C3:C10)		etc	..		..	..
13									
14					Simpson's diversity:			=1-H14	=1-I14

Figure 2.3 The spreadsheet showing the calculation of Simpson's index for two observations within the Wimbledon Common dataset. (This sheet shows the cell entries used. Circled cells should be copied and pasted up to their greyed-out boundaries.)

calculations, but helps keep the layout of the sheet clear so reduces the likelihood of mis-referencing or mispasting cells.) At the bottom of the two columns of data are cells which calculate the summed totals of the numbers above them. These will be used across the width of the sheet, and it is sufficient to enter the formula in the first cell (here B12) then

	A	B		D				
1		raw data			p_i			$\log_{10}p_i$	
2		Heath 1	Spoil 1	*	Heath 1		Spoil 1	*	Heath 1	Spoil 1
3	Achillea millefolia	0	10	*	0	0.1162791	*	0	-0.10866
4	Arrhenatherum elatius	0	70	*	0	0.8139535	*	0	-0.07277
5	Festuca rubra	80	0	*	0.727272727	0	*	-0.10058378	0
6	Calluna vulgaris	5	0	*	0.045454545	0	*	-0.06101921	0
7	Deschampsia flexuosa	25	0	*	0.227272727	0	*	-0.14623924	0
8	Heracleum sphondylium	0	0	*	0	0	*	0	0
9	Trifolium repens	0	5	*	0	0.0581395	*	0	-0.07183
10	Vicia sativa	0	1	*	0	0.0116279	*	0	-0.02249
12	sum:	110	86	0	1		1	0	-0.30784224	-0.27576
14									
15					Shannon diversity base 10:			0.30784224	0.275758
16					Shannon diversity base 2:			1.02262978	0.916047
17					Equitability			0.34087659	0.305349

Figure 2.4 The spreadsheet showing the calculation of Shannon's index for two observations within the Wimbledon Common dataset. (This sheet shows the actual numeric values.)

copy it across. The next stage is to calculate the proportion represented by each species, shown here in the p_i columns (columns E and F). Here each cell is a species value from the raw data divided by its appropriate column total. It is important to use the $ function to ensure that all values are divided by the same total. (If the cell E3 held the command '= B4/B13', its contents would be correct since cell E13 holds the column total, but when copied and pasted down to other cells the divisor would be different in each case. Instead the cell E3 holds the command '= B4/B$13', which may be pasted throughout the unshaded block from E3 to F10.) The column totals for these p_i columns should be checked to ensure that they equal 1.0.

These proportion values are now squared in columns H and I, giving the values of p_i^2 for each species, and these squared values are summed in cells H12 and I12. These are subtracted from 1.0 to give Simpsons' diversity indices.

A similar pair of spreadsheets is shown in Figures 2.4 and 2.5, generating the Shannon index (base 10 and base 2) and the equitability for the same pair of observations.

2.5.3 Example: The Liphook pine forest fungal dataset

Ecologists often record successional changes in community structure, such as in an initially bare habitat where colonisation starts with a few colonist species, followed by a gradual increase in numbers as new species arrive. An example of this is the Liphook pine forest fungal dataset (Table 1.9), and we can use this dataset to explore how community diversity indices change as the succession progresses.

The first year's data provide a good example of a low-diversity community. There were only two species recorded, and of the 281 individuals counted, 280 belonged to one species (the Deceiver *Laccaria proxima*). Thus the diversity of this observation is based on two p_i values, 280/281 and 1/281. These give diversity values of 0.007 for Simpson's D and 0.034 for Shannon's H. This process may be extended to give diversity values for each of the 35

	A	B	C	D	E	F	G	H	I	J
1		raw data			p_i			$p_i^* \log_{10} p_i$		
2		Heath 1	Spoil 1	*	Heath 1	Spoil 1	*	Heath 1	Spoil 1	*
3	Achillea millefolia	0	10	*	= B3/C$12	=C3/D$12	*	= IF(E3 = 0, 0, E3*LOG10(E3)	= IF(F3 = 0, 0, F3*LOG10(F3))	*
4	Arrhenatherum elatius	0	70	*	= B4/C$12	= C4/D$12	*	= IF(E4 = 0, 0, E4*LOG10(E4))	= IF(F4 = 0, 0, F4*LOG10(F4))	*
5	Festuca rubra	80	0	*	= B5/C$12	= C5/D$12	*	= IF(E5 = 0, 0, E5*LOG10(E5)')	= IF(F5 = 0, 0, F5*LOG10(F5))	*
6	Calluna vulgaris	5	0	*	etc..	..	*	etc..	..	*
7	Deschampsia flexuosa	25	0	*	..	..	*	..	..	*
8	Heracleum sphondylium	0	0	*	..	..	*	..	..	*
9	Trifolium repens	0	5	*	..	..	*	..	..	*
10	Vicia sativa	0	1	*	..	..	*	..	..	*
12	sum:	= SUM (B3:B10)	= SUM (C3:C10)		= SUM(E3: E10)	= SUM(F3:F10)		= SUM(H3:H10)	= SUM(I3:I10)	
14						Shannon diversity base 10:		= 0-H12	= 0-I12	
15						Shannon diversity base 2:		= H14/LOG10(2)	= I14/LOG10(2)	
16						Equitability		= H14/LOG10(8)	= I14/LOG10(8)	

Figure 2.5 The spreadsheet showing the calculation of Shannon's index for two observations within the Wimbledon Common dataset. This sheet shows the cell entries used. Circled cells should be copied and pasted up to their greyed-out boundaries. Note that the value 8 used in the equitability calculation is the number of species in the dataset. This could also be obtained by counting the number of names in column B.

Table 2.3 The diversity indices calculated for the Liphook pine forest data.

Year	Plot	Species richness	Shannon's index	Simpsons index	Year	Plot	Species richness	Shannon's index	Simpsons index
86	1	2	0.034	0.007	88	5	3	0.506	0.293
86	2	2	0.778	0.354	88	6	6	0.384	0.405
86	3	2	0.872	0.414	88	7	6	0.529	0.554
86	4	2	0.276	0.091	89	1	7	0.507	0.539
86	5	2	0.684	0.298	89	2	6	0.449	0.397
86	6	2	0.659	0.283	89	3	7	0.243	0.252
86	7	2	0.966	0.476	89	4	7	0.388	0.414
87	1	5	0.037	0.017	89	5	6	0.379	0.308
87	2	5	0.358	0.260	89	6	5	0.539	0.447
87	3	4	0.693	0.505	89	7	6	0.279	0.250
87	4	5	0.545	0.472	90	1	10	0.717	0.751
87	5	2	0.82	0.381	90	2	8	0.395	0.355
87	6	3	0.094	0.037	90	3	7	0.353	0.315
87	7	4	0.442	0.308	90	4	9	0.392	0.405
88	1	7	0.413	0.413	90	5	6	0.615	0.537
88	2	7	0.484	0.553	90	6	6	0.567	0.497
88	3	5	0.578	0.554	90	7	6	0.251	0.193
88	4	7	0.534	0.599					

observations, listed in Table 2.3. When graphed, there is a disappointing lack of a clear trend (Figure 2.6), rather a stabilising of values. This is due in part to the dominance of the dataset by one species *Suillus bovinus*, which led to low p_i values for all other species. (An extended version of this dataset did show a steady increase in diversity in later years,

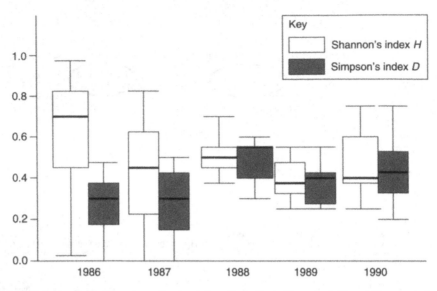

Figure 2.6 A Graph showing how the two diversity indices, Shannon's *H* and Simpson's *D*, alter with time in the Liphook pine forest fungal dataset. Boxes indicate the inter-quartile range, the tails represent maxima and minima and the thick dark lines the medians.

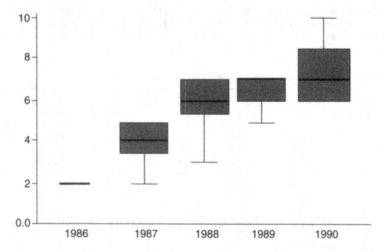

Figure 2.7 Graph showing how the species richness of the Liphook pine forest fungal dataset alters with time. Boxes indicate the inter-quartile range, the tails represent maxima and minima and the thick dark lines the medians.

coinciding with the decline of *Suillus bovinus*.) Here again, the only diversity index to show a clear pattern was the simplest of all, species richness, which increases steadily from year to year (Figure 2.7).

2.5.4 Example: The Alaskan streams dataset

The diversity indices describing the crustacean diversity are shown in Table 2.4, and are presented graphically in Figures 2.8–2.10. It can be seen that the community diversity

Table 2.4 Diversity indices for the Alaskan streams dataset.

site	rep	age, years	Simpson	Shannon	spp	site	rep	age, years	Simpson	Shannon	spp
stonefly	1	18	0	0	1	berg s	4	167	0.811	0.562	2
stonefly	2	18	0	0	1	berg s	5	167	0.96	1.055	3
stonefly	3	18			0	tyndall	1	147	0.639	0.886	4
stonefly	4	18			0	tyndall	2	147	0.771	0.534	2
stonefly	5	18			0	tyndall	3	147	0.341	0.375	3
wolf pt	1	48	0	0	1	tyndall	4	147	0.811	0.562	2
wolf pt	2	48	0	0	1	tyndall	5	147	0.998	0.691	2
wolf pt	3	48	0	0	1	rush pt	1	192	1	0.693	2
wolf pt	4	48	0	0	1	rush pt	2	192	0	0	1
wolf pt	5	48	0	0	1	rush pt	3	192	0	0	1
berg n	1	167	1	0.693	2	rush pt	4	192	0.921	1.011	3
berg n	2	167	0.865	0.95	3	rush pt	5	192	0	0	1
berg n	3	167	0.777	0.853	3	carolus	1	1377	0.65	0.451	2
berg n	4	167	0	0	1	carolus	2	1377	0.65	0.451	2
berg n	5	167	0.226	0.249	3	carolus	3	1377	0.593	0.822	4
berg s	1	167	0.946	1.311	4	carolus	4	1377	0.969	1.56	5
berg s	2	167	0	0	1	carolus	5	1377	0.876	1.215	4
berg s	3	167	0.865	0.95	3						

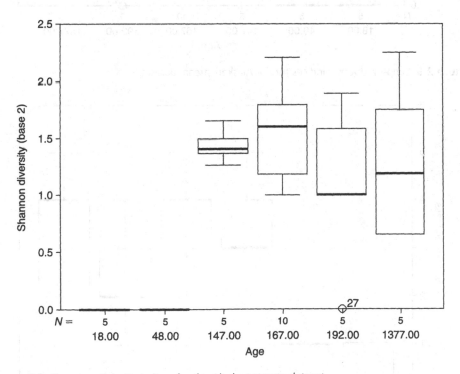

Figure 2.8 Shannon diversity indices for the Alaskan stream dataset.

remains zero (none or a single species only) until around 50 years, but that there is little difference between the diversities of mid-successional streams and the oldest site. The zero
values for diversity found in the youngest sites include samples with only one species in (a
true zero-diversity) and samples with no species at all (where species richness is truly zero

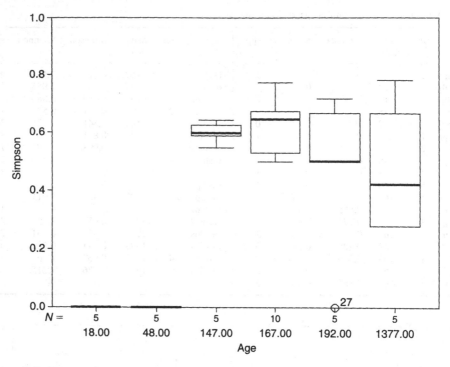

Figure 2.9 Simpson diversity indices for the Alaskan stream dataset.

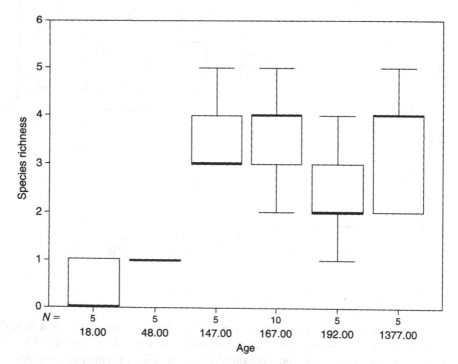

Figure 2.10 Species richness as a function of age for the Alaskan stream dataset.

but the derived indices are best treated as missing values). In this case simple calculations of diversity show a difference between the youngest communities and others but give little insight into the dramatic differences in community composition between the middle-aged and the oldest stream.

2.5.5 Example: Nutrient enrichment of Dutch chalk grasslands

This example comes from Bobbink (1991), and concerns the effects of increased atmospheric pollution on the growth of coarse grasses in Dutch chalk downlands. The root of the problem is the high levels of nitrogen depositing from atmospheric sources in the Netherlands, due mainly to ammonia release from intensive livestock rearing (van Breemen and van Dijk, 1988). Among other effects, this stimulates coarse grasses such as Tor grass *Brachypodium pinnatum* in preference to the rich community of low-growing, less vigorous herbs typical of unfertilised chalk soils (Bobbink and Willems, 1987).

Bobbink (1991) set up five experimental plots on areas of chalk downland in which *Brachypodium pinnatum* was present but not dominant, and divided each up into five sub-plots which received different combinations of nitrogen, phosphorus and potassium fertilisers. The results showed a significant increase in biomass in fertilised plots, and decreases in numbers of plant species. These data were summarised using the Shannon index and its derivative equitability, which were then analysed by ANOVA.

The results are shown in Figure 2.11, which show that both indices of diversity decreased significantly in the nitrogen-only treatment (although not in the nitrogen + phosphorus + potassium treatment). This was explained by the ability of *Brachypodium pinnatum* to flourish on high levels of nitrogen in conjunction with low levels of phosphorus, using its height to shade out other species thereby reducing the biodiversity and conservation value of the habitat.

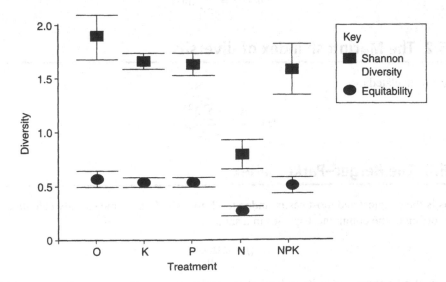

Figure 2.11 Diversity and equitability of chalk vegetation under five nutrient regimes (bars show 1 s.e.). Replotted from data in Bobbink (1991).

2.6 Further diversity indices

For completeness, further indices will be defined without detailed explanation or worked examples.

2.6.1 The Brillouin index

This is recommended as a substitute for the Shannon index where the randomness of sampling is not guaranteed, such as insect trapping (Southwood, 1978) where different species may be attracted to the light to different degrees. It is also suitable where the entire population has been completely censused.

It is calculated as:

$$HB = \frac{\ln(N!) - \sum \ln(n_i!)}{N}$$

where

HB = the Brillouin index
N = total number of individuals in the sample
n_i = number of individual of species i
$N!$ means the factorial of $N = 1 \times 2 \times 3 \times 4 \times \cdots \times N$
$\ln(x)$ = natural logarithm of x (or logarithm base e)

Unlike the indices described above, this index varies with sample size as well as with the relative proportions of species. It has the pragmatic limitation that the factorial function increases extremely rapidly, so that a scientific calculator will be unable to calculate the factorial of a number above 69. This can be bypassed by using the result that

$$Ln(N!) = \sum_{i=1}^{i=N} \ln(i)$$

2.6.2 The Macintosh index of diversity

This is calculated as:

$$D = \frac{N - \left(\sum n_i^2\right)^{1/2}}{N - N^{1/2}}$$

2.6.3 The Berger–Parker index

This is the simplest and most easily understood diversity index, since it only calculates the proportion of the commonest species in a sample:

$$d = \frac{N_{max}}{N}$$

It may also be expressed as its reciprocal, which has the advantage that low values indicate a low-diversity community and vice versa.

3

Linear regression: simple and multiple

3.1 Introduction

The techniques described in this chapter all involve fitting straight lines to data. This means plotting data as coordinates (points) in an imaginary space, then finding best-fit lines through these graphs. In this situation there are two distinct but linked questions that may be asked. The first is to ask for a precise definition of the relationship between the two variables. This is achieved by calculating an equation which defines the best-fit line, known as the **regression equation**. The second question is to establish how likely it is that the pattern observed in your data could be the result of random chance. This involves creating the null hypothesis of no association between the variables, and then calculating a **correlation coefficient**, from which a probability may be derived.

The simplest form of regression is the familiar two-variable (bivariate) approach, which is routinely taught in basic courses on statistical analysis, usually in conjunction with the Pearson correlation coefficient (symbol r) and Spearman's correlation coefficient (symbol r_s). The same approach can be used to fit straight-line (linear) models to datasets containing three or more variables, giving rise to the technique known as multiple linear regression. (As a point of detail, multiple linear regression finds best-fit planes in a multidimensional data space rather than the simple straight lines fitted to bivariate data.) As with diversity indices, multiple linear regression is not strictly a form of multivariate analysis because its output consists of a single equation (albeit one that may have several terms). This technique is included here because it is routinely used to explore or describe datasets which contain many variables, and because it forms part of truly multivariate techniques which will be encountered later in the book (Chapters 9 and 10).

This chapter will start off by explaining the notion of a data space, because this concept is crucial to an understanding of all regression techniques, along with most of the ordination techniques that come later in the book. This will be followed by an introduction to the different ways that best-fit lines can be defined (explaining why there is generally no single best-fit line through a dataset), and showing how one best-fit model leads to standard bivariate linear regression and correlation, in order that the deep similarities between these and multiple linear regression may readily be highlighted. Only then will the principles of multiple linear regression itself be set out, followed by examples and warnings of pitfalls for the unwary. The underlying algebra (which allows all forms of regression to be unified into one equation) will be supplied in Appendix 2 at the end of the book,

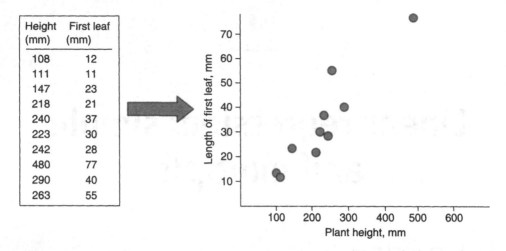

Height (mm)	First leaf (mm)
108	12
111	11
147	23
218	21
240	37
223	30
242	28
480	77
290	40
263	55

Figure 3.1 A scattergraph as a simple data space. Here an environmental data set (height and first leaf length of *Dactylorhiza* orchids) is shown both as numeric data and as a scattergraph. The scattergraph is in fact a portrayal of the data as coordinates in a two-dimensional data space.

but it will not be necessary to read this appendix to understand and apply multiple linear regression.

3.2 The data space

The key to understanding all forms of regression (indeed virtually all the techniques described in this book) is to think visually, to imagine data as describing patterns of points in space rather than as blocks of numbers. This involves a step of abstraction, creating the notion of a space which has no physical reality so cannot be seen or measured. In fact this is no more abstract than plotting a graph on paper – a familiar exercise in which two sets of numbers are converted into a new two-dimensional pattern. Figure 3.1 gives an example of a dataset (height and leaf length of marsh orchids *Dactylorhiza* spp from an ex-industrial site in Lancashire) portrayed as a set of coordinates in a scattergraph. This graphing is precisely the same as treating the data points as locations in a two-dimensional data space. The space has no tangible reality – one axis shows the height of orchids, the other the length of their first leaf – but providing the axes are at 90 degrees to each other (the state when the axes are said to be orthogonal to each other) each point in the space corresponds to a unique combination of measurements. This is a two-dimensional data space.

This principle may be extended to include more than two variables. Figure 3.2 shows how a third variable (in this case the number of flowers) may be added to the original data, requiring a three-dimensional space to represent the information completely. This defines a three-dimensional data space. Such a data space cannot be reproduced exactly in a two-dimensional portrayal (such as an image on paper), but a good idea of the pattern may be obtained by examining several different perspectives of the same image. Computer packages routinely offer the option of viewing and rotating three-dimensional graphs, and it is easier to appreciate the shape of such a dataset by interactively rotating its image on a screen than it is by examining static images on paper.

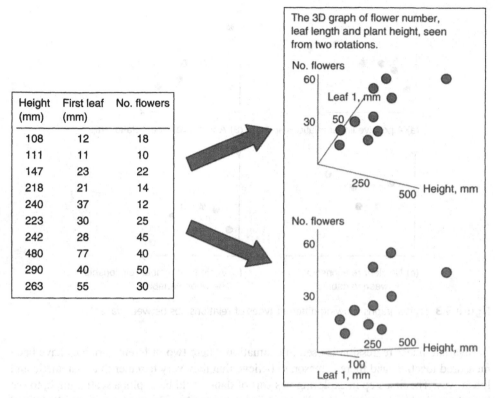

Height (mm)	First leaf (mm)	No. flowers
108	12	18
111	11	10
147	23	22
218	21	14
240	37	12
223	30	25
242	28	45
480	77	40
290	40	50
263	55	30

Figure 3.2 A three-dimensional data space. We now introduce a third variable to the dataset – in this case the number of flowers on each orchid. The three variables now define a three-dimensional data space, which can be visualised using three-dimensional graphs.

It can be seen that this same procedure could, in principle, be used to define four-dimensional data spaces (using four variables), five-dimensional, and so forth. It will also be realised that visualising data spaces with four or more dimensions is simply not feasible without some method of collapsing the shape into two-dimensional images. The idea of using two-dimensional representations to examine the shape of higher dimensional objects will be a recurring theme throughout multivariate analysis.

3.3 Bivariate linear regression

3.3.1 Basics

Unlike the majority of this book, bivariate linear regression is a standard component of all undergraduate statistics courses. The subject will be introduced here only because its fundamental concepts are easily extended to cover multiple linear regression. Nowadays access to software that performs bivariate linear regression analysis is nearly universal, even being built into cheap pocket calculators. The explanations given here will therefore focus on the underlying principles without going into the depth available in standard statistics textbooks.

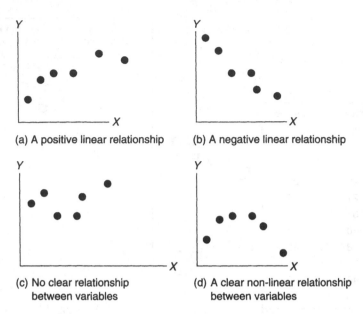

(a) A positive linear relationship   (b) A negative linear relationship

(c) No clear relationship   (d) A clear non-linear relationship
    between variables   between variables

Figure 3.3 Scattergraphs showing different types of relationships between variables.

Bivariate linear regression is used in a situation where two different variables have been measured together, and there is reason to believe that they vary together in a systematic and linear way. The first step in exploring this sort of data would be to plot a scattergraph, to see how the variables change together. Figure 3.3 gives examples of scattergraphs; graphs (a) and (b) suggest a linear relationship between the variables and would certainly invite a full linear regression. Graph (c) appears to show no relationship between the data (although this could be tested), while (d) shows a systematic but curved trend that would require some form of non-linear analysis (a complex subject which will not be explored further in this book).

There is an important consideration about the data space used in any regression analysis (bivariate or multiple), which separates out regression analysis from many apparently similar multivariate techniques introduced later in the book. All forms of linear regression assume that one of the variable is dependent on the other(s). This is known as the **dependent variable**, while the other variable (assumed to be driving the system) is known as the **independent variable**. By convention the dependent variable is plotted on the vertical axis of a graph (the Y-axis) while the independent variable goes on the horizontal axis (the X-axis). In the case of multiple linear regression there may be many 'independent' (or properly 'explanatory') variables, but still there is only one dependent variable.

Generally it is possible to identify which variable is the dependent and which is the independent. By way of example, consider the case of a graph relating energy output of a wind turbine to the ambient wind speed. In this case the wind is turning the windmill not vice versa, so that energy output is the dependent variable so is plotted on the Y-axis, while wind speed is the independent variable. (This example could be turned on its head for a wind tunnel in which a fan turns to produce an air movement – in this case wind speed would be dependent on energy consumed by the fan.) Table 3.1 gives some examples of types of environmental datasets in which the distinction between dependent and independent variables is clear.

Table 3.1 Examples of common types of environmental datasets in which bivariate linear regression may be suitable and a clear distinction exists between the dependent and independent variable.

Nature of the dataset	Independent variable (the X-axis)
Any time series	Time is always the independent variable
Impact of a disturbance (trampling, grazing, etc.) on a system	Intensity of disturbance
Measurements along a transect	Distance along the transect
Addition of measured amounts of treatments (fertilisers, pesticides, drugs, etc.) to an experimental system (living or inanimate)	Dose of treatment added
Hydrological data for catchments	Physical features of the catchment (slopes, channel width, etc.)
Vertical profiles in litter/soil	Depth in the profile
Allometric studies (exploring how properties scale with body size)	Body mass

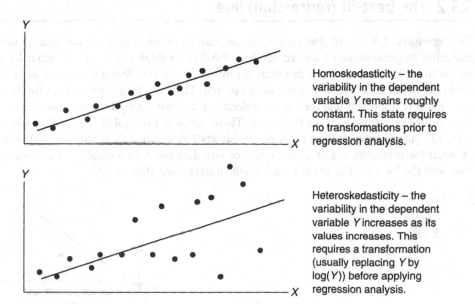

Homoskedasticity – the variability in the dependent variable Y remains roughly constant. This state requires no transformations prior to regression analysis.

Heteroskedasticity – the variability in the dependent variable Y increases as its values increases. This requires a transformation (usually replacing Y by $\log(Y)$) before applying regression analysis.

Figure 3.4 Examples of heteroskedasticity and homoskedasticity.

The distinction between dependent and independent variables is essentially a philosophical question, but is important in a practical sense because it affects the subsequent calculations. The precise form of the relationship between two variables is not symmetrical: a different line will be obtained if X is assumed to be dependent on Y from that calculated assuming Y to depend on X. This is because the standard linear regression model makes some assumptions about the data, one of which is that the independent (X) variable is measured perfectly without error, and that all the measurement errors occur within the dependent variable. (In practice this assumption is often violated.) Other assumptions are that the data are drawn from a normal distribution, and that the variance of data remains constant over its entire distribution. This latter condition is known as homoskedasticity; its opposite is heteroskedasticity, which can be seen graphically in Figure 3.4, where it may be seen that the variation in the dependent variable becomes greater as its values increase. Data suffering from heteroskedasticity should be transformed (usually by log transformation) to stabilise its variance prior to analysis.

Situations often arise in which there is no clear distinction between the dependent and independent variable. An example of this is shown in Figure 3.2, where the choice of Y-axis is rather arbitrary. Another common situation where this might arise involves data about environmental chemical measurements, such as indices of water quality. Within such data any two chemicals could be found to be correlated, but it would be difficult to justify asserting that either chemical were dependent on the other. In such instances the standard model of linear regression should be used with caution, and alternative models (such as a reduced major axis – see Figure 3.8) may be more suitable. It is however still legitimate to calculate a significance value using correlation analysis since (unlike regression) this does not require that a distinction be drawn between dependent and independent variables.

3.3.2 The best-fit (regression) line

The definition of a **best-fit line** involves mathematical considerations of the data (in contradiction to generations of undergraduates who like to think that it means 'what looks alright to me'!). The problem is that there are infinitely many different lines that can be fitted to any set of data – but only one can be optimal. There are formal procedures for doing this, and it is important that the reader understand the philosophy behind these, although the resulting equations need not be learned. There are in fact several different definitions of 'best fit' but all involve the notion of '**residuals**'. A residual is simply the difference between the predicted and observed value for any data point on a graph – an estimate of how well the 'best fit' line fits to each individual data point (Figure 3.5).

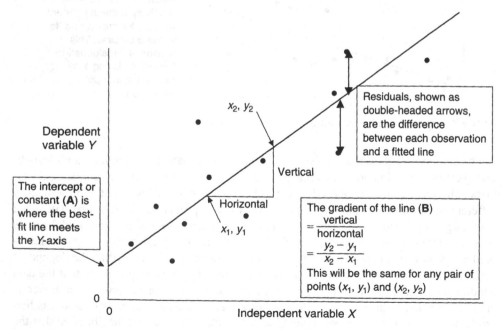

Figure 3.5 The meaning of the gradient, intercept and residuals of a best-fit line.

The best-fit line is always defined in terms of its residuals. The better the fit, the smaller are the residuals. In essence the definition of a best-fit line involves aggregating information from all the residuals in order to find the smallest possible set of residuals. But how to aggregate the information from all your residuals into one number? It might seem intuitively simply to add the residuals up, but in fact the mathematics turn out to work better if you add up the squared values of the residuals. Best-fit lines are almost invariably defined as minimising the sum of residuals squared – this is known as the least squares fit. Given this information, it is possible to derive best-fit relationships between two variables, expressed as an equation.

Any straight line in two-dimensional (X-Y) space can be written by an equation of the form:

$$Y = A + B \times X \qquad (3.1)$$

where Y is the dependent variable, X is the independent variable, and A and B are numbers which are constant for any given line. The constant A is the value of Y when X is 0 (because at this point $Y = A + B \times 0 = A + 0 = A$ for any value of X), and is known as the **intercept**. B is called the **regression coefficient**, which is actually the **gradient** of the line, meaning that when X increases by 1, Y increases by B. (A frequent but older version is to write equation 3.1 as $Y = M \times X + C$, which is clearly identical except for the symbols for the gradient and intercept.) These symbols are explained visually in Figure 3.5.

The regression equation for a bivariate dataset simply involves finding best estimates for A and B. This is usually calculated by a software package, using the equations given in Figure 3.6. However it is important to stress that there is no one unique line of best fit, rather a choice from several options, depending what constraints one chooses to impose. The default (indeed often the only option available in computer packages) follows the diagram shown in Figure 3.6: the residuals all stand vertically, and the best-fit line passes

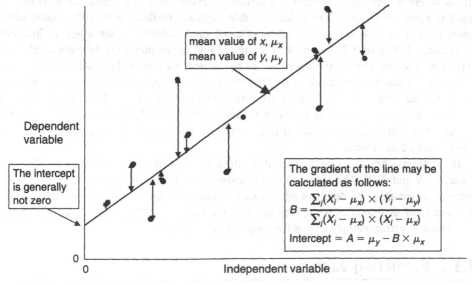

Figure 3.6 The standard model for a best-fit line: Residuals are vertical and the line passes through the overall mean of the data. The significance of this relationship may be measured by calculating a correlation coefficient.

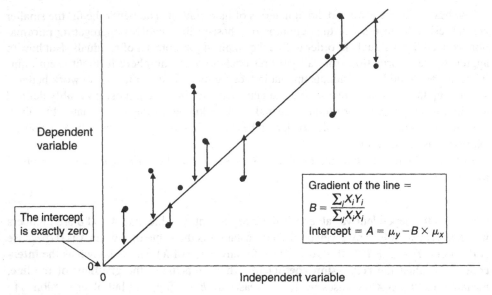

Figure 3.7 The zero-intercept model for a best-fit line: Residuals are vertical and the line passes through (0, 0). No significance testing or correlation coefficient is possible for this model.

through the overall mean of the dataset (the point whose X-coordinate is the mean value of all the X-axis data, and whose Y-coordinate is the mean of all Y-axis data). This model allows the calculation of correlation coefficients and estimates of significance (see below).

It is often forgotten that there are other models available: one example is that the line may be constrained to pass through the origin of the graph (0, 0). This is very useful for situations where a zero value for the independent variable may confidently be expected to result in a zero value for the dependent variable, such as a calibration curve in which colour intensity is plotted against concentration of a coloured material. Again a best-fit line may be calculated (Figure 3.7), but under this model no significance may be calculated. (The explanation for why no significance may be calculated is given in Appendix 2.)

Finally, in cases where there is no clear dependent variable it is possible to calculate a best-fit line using residuals which are orthogonal to it (Figure 3.8). This is known as a reduced major-axis line, and has been favoured for allometric work (Sokal, 1995). It is in fact a two-dimensional version of principal components analysis (Chapter 6), although rarely referred to as such.

These approaches are compared by fitting different best-fit lines to one dataset, shown in Figures 3.9 and 3.10. The standard and zero-intercept lines are almost identical, but the reduced major-axis line does not agree exactly, and if the lines were extrapolated outside the data their divergences would rapidly become sizeable. (This is one reason why one should avoid extrapolating outside the range of one's data.)

3.3.3 Predicting values

The calculation of a best-fit line allows a researcher to predict the value of the dependent variable for any value of the independent variable. An example is given in Figure 3.11,

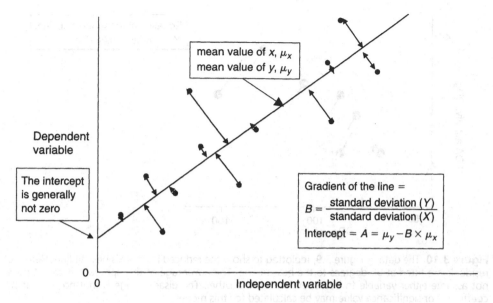

Figure 3.8 The reduced major-axis model for a best-fit line: Residuals are orthogonal and the line passes through the overall mean of the data. No significance testing is possible for this model.

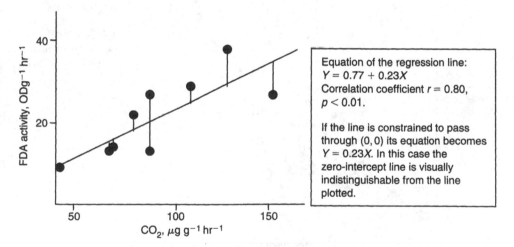

Figure 3.9 A graph of FDA activity against CO_2 respired for plant litters. These are the distances that each point lies away from a fitted line, and as such are an index of how well the line fits to the data. Here we see a graph showing a linear relationship between enzyme activity in plant litters as measured by the fluorescein-diacetate (FDA) assay and the amount of CO_2 respired (Stubberfield and Shaw, 1990), along with the vertical residual for each data point.

using a dataset relating erosion depth on marble tombstones against their age. These data came from a survey of three churchyards in the Sheffield area, and relied on the fact that lead letters are embedded into marble tombstones flush with the surface. As the stone is dissolved away by the acidity of rainwater, these letters gradually protrude from the surface, allowing an accurate estimate of the degree of erosion that has occurred. The data

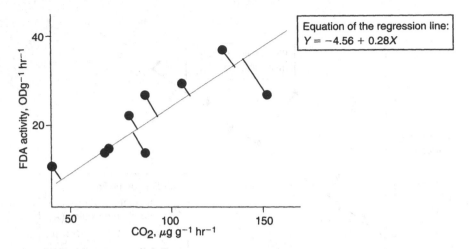

Equation of the regression line:
$Y = -4.56 + 0.28X$

Figure 3.10 The data of Figure 3.9, replotted to show the reduced major-axis best-fit line. Here the residuals are fitted at 90 degrees to the best-fit line. The advantage of this approach is that it does not assume either variable to be dependent on the other. The disadvantage is that no correlation coefficient or significance value may be calculated for this model.

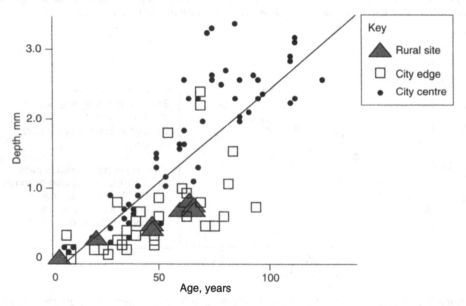

Key

▲ Rural site

☐ City edge

● City centre

Figure 3.11 Erosion depth on marble tombstones in three Sheffield churchyards as a function of the age of the stone. Basic regression equation: $Y = -0.03 + 0.027X$.

shown in Figure 3.11 clearly indicate that erosion depth increases with increasing age of the stone, suggesting the need for linear regression. (To a good approximation, this dataset meets the criterion that the independent variable be measured without error, since the age of the stone is given reasonably accurately by the first recorded date of death). Figure 3.11 gives the regression equation between erosion depth and age as:

$$\text{Depth (mm)} = -0.03 + 0.027 \times \text{age (years)}$$

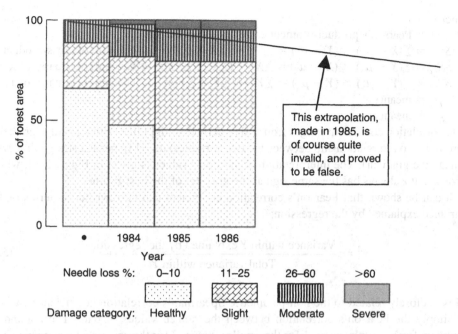

Figure 3.12 A warning about extrapolating outside the range of one's data. After the 1984 German forest dieback survey, extrapolations were made predicting further decline, with a total loss of forest cover around 2000. (This never happened.)

leading to the prediction that a 40-year-old stone should have an erosion depth of $(0.027 \times 40 - 0.03) = 1.05\,\mathrm{mm}$.

A general rule for when extrapolating using linear regression is that one should avoid extrapolating outside one's data. In the case of the marble erosion data used here it would be unwise to extrapolate erosion depths into the future, since local conditions may well be different in the future to historic levels (especially with regards to corrosive pollutants in the air). A cautionary example here comes from the German forest dieback of the mid-1980s (Shaw, 2000), where a forest health (based on needle loss) was found to have declined substantially between 1983 and 1984. This rate of loss was extrapolated (in early 1985) to a prediction that most of Germany's forests would be dead or dying by 2000. In fact there was been almost no further change in forest health (Figure 3.12), because the dieback was restricted to certain particularly base-poor soil types. The difference between 1983 and 1984 may have been an artefact caused by different measurement protocols.

3.3.4 The correlation coefficient

There are several different approaches to calculating a correlation coefficient, but the commonly used and most relevant to this book is **Pearson's product-moment correlation coefficient**, which is supplied as a default by all software packages and pocket calculators. For bivariate (X–Y) data this correlation coefficient is given by the formula:

$$r = \frac{SS_{xy}}{(SS_{xx} \times SS_{yy})^{1/2}} \tag{3.2}$$

where
r = Pearson's product-moment correlation coefficient
$SS_{xy} = \Sigma(X - \mu_x) \times (Y - \mu_y) = \Sigma XY - (\Sigma X \times \Sigma Y)/n$ ('The sum of cross-products')
$SS_{xx} = \Sigma(X - \mu_x) \times (X - \mu_x) = \Sigma X^2 - (\Sigma X \times \Sigma X)/n$ ('The sum of squares of X')
$SS_{yy} = \Sigma(Y - \mu_y) \times (Y - \mu_y) = \Sigma Y^2 - (\Sigma Y \times \Sigma Y)/n$ ('The sum of squares of Y')
$\mu_y = \text{mean}(y)$
$\mu_x = \text{mean}(x)$

The correlation coefficient ranges from 1.0 (a perfect straight line with a positive gradient) through zero (no relationship between the two variables) to -1 (a perfect straight line with a negative gradient). The interpretation of various r values is given in Figure 3.13, which shows that r always has the same sign as the gradient of the best-fit line.

It can be shown that Pearson's correlation coefficient can be rewritten in terms of the variance explained by the regression:

$$r^2 = \frac{\text{Variance within } Y \text{ explained by the regression}}{\text{Total variance within } Y} \qquad (3.3)$$

This is closely related to the non-parametric Spearman's correlation coefficient r_s, which is simply the Pearson's correlation between the ranked values of data. The column of numbers for X is replaced by 1 for the smallest value, 2 for the second smallest value, etc. up to N for the largest value (where N is the number of observations in the dataset). This exercise is repeated for the Y variable, and the correlation between these two new variables is calculated as above. An alternative expression is generally taught to students, given below as equation 3.6, which speeds the calculation but obscures the mathematical relationship!

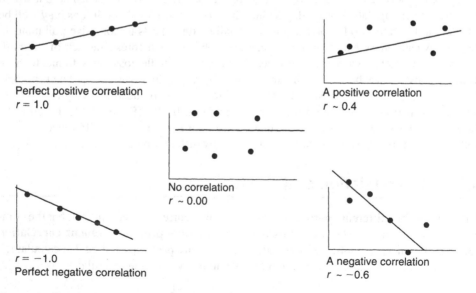

Figure 3.13 Visual explanations of values of Pearson's correlation coefficient: Its values ranges between 1.0 and -1.0: never >1 nor <-1.

Spearman's correlation coefficient $r_s = 1 - \dfrac{6 \times \sum D_i^2}{(N-1) \times N \times (N+1)}$ (3.4)

where $\mathbf{D}_i$ is the difference between the rank of the ith value of X and of the ith value of Y.

In principle any of the techniques in this book could be repeated on ranked data, but in practice this rather drastic rescaling of data is rarely used in multivariate analyses except for NMDS (Chapter 10).

The power of calculating a correlation coefficient is that it can be equated with a probability value. This is usually supplied by a statistics package (though not currently on pocket calculators), or can be obtained from standard statistical tables. For the value to be dependable, Pearson's correlation coefficient assumes that the data are derived from a normal distribution. (When the assumption of normality fails, it is necessary to find a transformation of the data to make them approximate to a normal distribution, or failing that, to replace Pearson's correlation coefficient by the non-parametric Spearman's correlation coefficient.) In order to find the significance of a correlation coefficient from standard tables it is necessary to find the number of degrees of freedom of the r value. This leads to a common minor slip, since degrees of freedom are usually calculated as $n - 1$, but for a bivariate regression they are given by:

$$df = n - 2$$ (3.5)

where
    df = the degrees of freedom
    n = the number of data points.
The notion of degrees of freedom sometimes causes conceptual problems, but can be explained simply in this case as an indicator of the scope for a fitted line to wobble. A graph with two data points has zero degrees of freedom – this means that a perfect line can always be fitted. It takes the inclusion of a third data point to give the possibility that the best-fit line will not fit the data perfectly (Figure 3.14). The same logic can be extended up one dimension to show that a three-legged seat cannot wobble, since three points in a three-dimensional space define a flat plane.

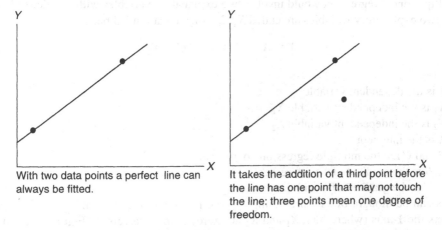

With two data points a perfect line can always be fitted.

It takes the addition of a third point before the line has one point that may not touch the line: three points mean one degree of freedom.

Figure 3.14 Visual explanations of why fitting a line removes two degrees of freedom.

3.4 Additional independent variables

3.4.1 Multiple linear regression

It is common that environmental datasets may contain more than one variable which may explain some result, and researchers are often faced with the need to decide between several possible explanations which need not be mutually exclusive. Examples of this might be questions such as whether plant colonisation of a bare site is better explained by soil nitrogen or water content? Is the degree of branching of stream systems better explained by geology, slope steepness or rainfall patterns? Does the composition of assemblages of spiders depend on vegetation structure or on its species composition? In such cases the 'independent' variables tend not to be independent of each other, and it is better to refer to them as explanatory variables (since they may explain some of the behaviour of the dependent variable).

One approach to this situation is to examine each of the explanatory variables in turn, in order to discover which has the most significant correlation with the dependent variable. This approach is valid, and in fact avoids some of the pitfalls inherent in multiple regression, but has disadvantages too. It requires the researcher to perform multiple tests, and can give unsatisfactory results when two or more factors independently affect the dependent variable. The usual solution to coping with multiple independent variables is to incorporate them all into the regression simultaneously. This is the procedure known as **multiple linear regression**, and in essence is identical to bivariate linear regression described above.

Bivariate linear regression works in a two-dimensional space (so can be shown on a standard graph) – multiple linear regression works in higher dimensional spaces, so is generally impossible to portray graphically. In both cases there must be only one dependent variable (Y) but multiple linear regression can have an indefinite number of explanatory variables, providing that there are sufficient data points. (There is a multivariate extension of this technique which includes multiple dependent variables as well as explanatory variables – this is canonical correlation analysis, described briefly in Chapter 10.)

The actual mathematics of multiple linear regression are rather more laborious than those for bivariate data and will invariably be performed on a computer, but the underlying model is identical except for the number of independent variables. The simplest case of multiple linear regression would involve two explanatory variables with one dependent. If the two explanatory variables are called X_1, X_2, then equation 3.3 becomes:

$$Y = A + B \times X_1 + C \times X_2 \tag{3.6}$$

where:
  Y is the dependent variable
  X_1 is the independent variable X_1
  X_2 is the independent variable X_2
  A is the intercept
  B and C are the multiple regression coefficients:
    B is the gradient of Y with respect to X_1
    C is the gradient of Y with respect to X_2.

This can be thought of as representing a plane in a three-dimensional system. The plane meets the Y-axis (where both X_1 and X_2 are zero) at the intercept A (Figure 3.15). If this plane is sliced in the X_1–Y plane the result would be a graph of a straight line with intercept

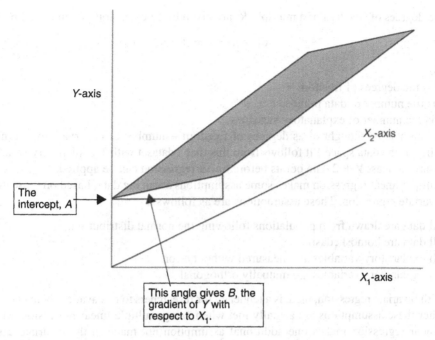

Figure 3.15 A three-dimensional graph showing the fitting of a plane (the 3D equivalent of a straight line in 2D space).

A and gradient B, and the same procedure in the X_2–Y plane would produce a graph of a line with intercept A and gradient C (Figure 3.15).

A more general version of equation 3.6 is applicable to any number of explanatory variables:

$$Y_i = A + \sum_{j=1}^{j=V} M_j \times X_{ji} \qquad (3.7)$$

where
Y_i is the dependent variable
X_j is the jth explanatory variable
A is the intercept
M_j is the jth regression coefficient
V is the number of explanatory variables.

The correlation between Y and all the explanatory variables is given by the R^2 value, which is a variance ratio already defined in equation 3.5. In the bivariate situation the square root of this variance ratio (r) tends to be used, since its sign indicates whether the gradient of the line is positive or negative. This is not feasible in multiple linear regression since there are many possible combinations of positive and negative gradients. Even in the simple trivariate case given in equation 3.7 there are four possible combinations of gradients: ($B > 0$ and $C > 0$), ($B > 0$ and $C < 0$), ($B < 0$ and $C > 0$), and ($B < 0$ and $C < 0$). This complexity cannot be summarised by r being positive or negative, so r in multiple regressions is replaced by its square.

The degrees of freedom of a multiple R^2 are given by an extension of equation 3.6:

$$df = n - V - 1 \tag{3.8}$$

where

　df is the degrees of freedom

　n is the number of data points

　V is the number of explanatory variables.

(This can also be thought of as degrees of freedom = number of observations − dimensionality of the data space.) It follows from this that a dataset with V explanatory variables must have at least $V + 2$ data points before linear regression can be applied.

Multiple linear regression makes some assumptions about the data, based on those made by bivariate regression. These assumptions are as follows:

1. All data are drawn from populations following the normal distribution.
2. All data are homoskedastic.
3. All explanatory variables are measured without error.
4. All explanatory variables are mutually orthogonal.

As with bivariate regression, there is a tendency for researchers to be cavalier about checking whether these assumptions are actually met when using multiple linear regression. Multiple linear regression makes one additional assumption not made in the bivariate case − number 4, that the explanatory variables are orthogonal. This means that they contain a balanced mixture of values, just as a properly designed experiment should contain a balanced mixture of all combinations of treatments. Formally it means that the correlations between each pair of explanatory variable approximates to zero. A frequent problem with multiple linear regression occurs when the explanatory data are not orthogonal but instead show some degree of co-correlation. This condition is known as collinearity, and can make the regression difficult or misleading to interpret.

3.4.2 Stepwise multiple linear regression

Multiple linear regression is a standard feature on most statistical packages, and usually comes with several options concerning entry of data into the regression. The approach described above (in which the dependent variable is regressed onto all the explanatory variables simultaneously) is generally the default, but often **stepwise multiple linear regression** is also offered. This involves entering explanatory variables into the regression one at a time, in such a way that each variable entered gives the best possible increase in R^2 value. This procedure is explained in Table 3.2.

It should be noted that each explanatory variable added into the regression equation will increase the R^2 value, or at worst will leave it unchanged. Adding an explanatory variable cannot decrease the R^2 value. This can be understood by considering a simple case with two explanatory variables, X and Z and one dependent variable Y. Step one of any stepwise multiple regression involves identifying which of the explanatory variables has the highest R^2 value when regressed alone with the dependent variable. Let us assume that in this hypothetical example X has the higher R^2 value with Y. This generates the first regression equation:

$$Y = A + B \times X, \text{ with a corresponding } R^2 \text{ value} \tag{3.9}$$

Table 3.2 The flowchart for performing stepwise multiple regression. It should be emphasised that this would normally be performed by a computer package.

---

1. Set $N = 0$.

2. Set up the regression equation with N explanatory variables, and find its R^2 value. (When $N = 0$ this starts off at $R^2 = 0$.)

3. For each unused variable in turn, calculate the regression equation between it and the dependent variable and calculate the resulting R^2 value.

4. Increment N by 1.

5. Identify the independent variable which gives the highest new R^2 value, and call this V_n.

6. Calculate the F ratio corresponding to the inclusion of V_n in the model.

7. If the F value calculated in step 6 is statistically significant, enter V_n into the regression equation as the Nth term and return to step 3.

8. When no more terms are statistically significant the analysis usually terminates, although in theory all remaining independent variables can be entered into the regression.

On the second step the remaining explanatory variable is entered:

$$Y = A + B \times X + C \times Z, \text{ with a new } R^2 \text{ value} \tag{3.10}$$

Notice that if C in equation 3.10 takes the value of 0, the equation then becomes identical to equation 3.9, with the same R^2 value. This proves that adding Z into the regression equation cannot reduce the overall R^2 value, since there is at least one regression equation that includes Z but which gives the same R^2 value as before.

Since adding a new variable (even random noise) into any multiple linear regression is almost certain to produce a slight increase in the overall R^2 value, it is valid to ask whether the change is meaningful. Fortunately it is possible to calculate a variance ratio (or F value) corresponding to the addition of a new variable, which then allows a probability to be calculated. Most packages that implement stepwise multiple linear regression will terminate the analysis when no variables remain which produce a significant improvement in the R^2 value. This can be a useful filter, identifying which variables are actually important. It would of course be possible to continue the analysis until all variables have been entered.

A less common alternative is **backwards stepwise multiple linear regression**. This mirrors stepwise multiple linear regression, but involves progressive removal of variables instead of their addition. The first step in a backwards multiple linear regression involves entering all the explanatory variables into the regression equation simultaneously. The second step is to identify the explanatory variable whose **removal** from the regression equation causes the smallest decrease in R^2. This procedure is usually repeated until the removal of a variable causes a significant decrease in the R^2 value, although in theory one could continue until only one explanatory variable remains. Curiously, when a full model is deconstructed this way the results may sometimes differ from the findings of a forward stepwise regression. Because of this sort of unreliability, backwards stepwise multiple linear regression is used by researchers less than forward stepwise, although it may be used as a confirmatory step: if forward and backward stepwise regressions arrive at the same set of explanatory variables one can be confident that these are indeed the important set to consider.

3.5 Pitfalls for the unwary

The wide popularity of multiple linear regression as a statistical tool belies the fact that there are a number of serious problems in the application of this technique to environmental data. These arise from the assumptions inherent in the model underlying multiple linear regression (listed above under 3.4.1).

The first three assumptions relate to the distributions of the variables, and if violated will make the calculated significance levels unreliable. Assumptions 1 and 2 (normality and homoskedasticity) can be checked by visual inspection of the data backed up by standard tests, and controlled by suitable transformations of the data. Assumption 3 (absence of error in the explanatory variables) cannot be controlled, and is often quietly ignored by environmental researchers! In many cases it might be preferable to replace multiple linear regression by principal components analysis (Chapter 6), which does not make such restrictive assumptions.

The fourth assumption (orthogonality of explanatory variables) probably causes more problems than any other in multiple linear regression, since its effects can dramatically distort the interpretation of output from an analysis. It is most unusual to find an environmental dataset in which all the variables are anywhere near to being orthogonal – a high degree of collinearity is the norm. This does not affect the R^2 value generated by a regression analysis, nor the significance test, nor the reliability of predicted values, but it does make the regression coefficients potentially misleading.

In bivariate linear regression the interpretation of regression coefficients is simple: if the correlation is significant and the regression coefficient is positive, an increase in the independent variable (X) is associated with an increase in the dependent variable (Y). Similarly, a negative regression coefficient means that an increase in X is associated with a decrease in Y. This simple pattern should also hold true for multiple linear regression, but only if the explanatory variables are orthogonal. When data are collinear, it becomes difficult if not impossible to interpret regression coefficients (Montgomery and Peck, 1982). It would be possible to have a dataset with one dependent variable (Y) and two explanatory variables (X_1 and X_2), in which Y is correlated positively both with X_1 alone and with X_2 alone, but negatively with X_2 when entered into a multiple regression. The R^2 value and the predicted values would be correct and easy to interpret, but the gradients (although technically correct) would be very misleading.

An interesting approach to solving the problem is given by Gardiner (1979), who was interested in using a set of six independent geographical variables to predict drainage density in catchments. The independent variables were highly collinear, so he used an ordination technique called principal components analysis (Chapter 6) to generate six perfectly orthogonal new variables, which were then entered into a multiple linear regression.

3.6 Worked examples

3.6.1 Example

This first example comes from the marble erosion dataset first introduced in Figure 3.11. This concerns 100 separate values of erosion depths on marble tombstones of various ages at three churchyards around Sheffield. In Section 3.3 these data were simply lumped

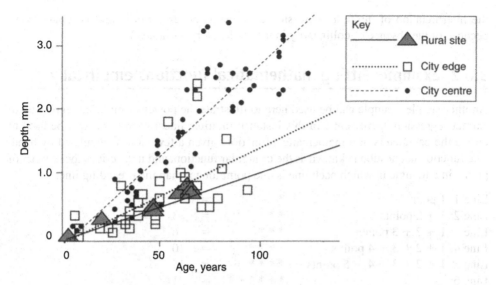

Figure 3.16 Erosion depth on marble tombstones in three Sheffield churchyards as a function of the age of the stone. A separate regression line has been fitted for each of the three churchyards.

together and analysed as one block, but in fact the three churchyards were at different locations and show signs of differing rates of erosion. Churchyard number one was in the city centre, churchyard two was on the outskirts of the city, while number three was in the moors well upwind of the polluted city centre. It can be seen that when a separate regression line is calculated for each site, the city-centre site suffered the highest rates of erosion while stones in the clean air of site three experienced the lowest rates of erosion (Figure 3.16). In this case the assumptions underlying multiple linear regression are reasonably well met since the data distributions are suitable, and the explanatory variables (stone age and distance from city centre) are both measured accurately and are approximately orthogonal (since a similar spread of ages was measured at each site) – there is little collinearity in the data.

These data can be examined using multiple linear regression by including distance from the city centre in the regression as well as the age of the tombstone. (An alternative approach would be to compare the gradients using analysis of covariance: the model for this is explained in Appendix 2.)

Analysis by a standard software package gave the multiple regression equation linking erosion depth (in mm) to age of tombstone (years) and distance from city centre (km) as:

$$\text{erosion} = 0.42 + 0.024 \times \text{age} + (-0.13) \times \text{distance},$$
$$R^2 = 0.73 \text{ (with 97 degrees of freedom)}$$

which can be compared with the bivariate equation:

$$\text{erosion} = -0.28 + 0.027 \times \text{age}, \quad R^2 = 0.63 \text{ (with 98 degrees of freedom)}$$

It will be seen that the multiple regression has explained more of the total variance in the data than the bivariate regression (73% compared to 63%), and left the gradient with respect to the age of the tombstone almost unchanged at 2.4 mm per century. The gradient of erosion with respect to distance from the city centre is negative, suggesting that churchyards at greater distances experience less erosion. (It should be emphasised however that the interpretation of multiple regression coefficients needs caution, and is reliable here because of the absence of collinearity in the explanatory variables.)

3.6.2 Example: Fitting mathematical functions empirically

Another simple example can be used here to illustrate the power of multiple regression to extract regression coefficients. In this instance multiple linear regression will be used to derive the parameters of a mathematical function, given a few values calculated by hand. The function in question is known as the triangular function, and it describes the number of points in a triangle in which each line is one point longer than the preceding line:

Line 1: 1 point	*	=	1
Line 2: 1 + 2 points	* *	=	3
Line 3: 1 + 2 + 3 points	* * *	=	6
Line 4: 1 + 2 + 3 + 4 points	* * * *	=	10
Line 5: 1 + 2 + 3 + 4 + 5 points	* * * * *	=	15
Line 6:	* * * * * *	=	21

It is possible to write the triangular function as a polynomial equation, meaning one written as

$$F(x) = A + B \times X + C \times X^2 + D \times X^3 \cdots,$$

and multiple linear regression can be used to estimate the values for A, B, C, D etc. The data in Table 3.3 consist of one dependent variable (the actual value of the triangular function) and three explanatory variables (values of x, x^2 and x^3). A standard statistical package provided estimates of 8×10^{-14} for the constant A, 0.5 for B, 0.5 for C, and 5×10^{-15} for D. The values for A and D are minute, and in fact approximate to zero (with small arithmetic errors having arisen during calculation). Ignoring these tiny values, Table 3.3 shows the well-known result that the triangular function may be written as:

$$f(x) = 0.5 \times x + 0.5 \times x^2$$

It should also be noted that the R^2 value given in Table 3.3 is 1.0; in other words the data are a perfect fit. This is expected when fitting parameters to mathematical functions, but is almost unknown when handling real environmental data.

3.6.3 Example: Simple discriminant functions

Another application of multiple linear regression involves the creation of **discriminant functions**. These are calculations that allow a researcher to allocate an individual into predefined categories, based solely on numeric measurements. The idea dates back to Fisher (1936, 1940). There are several different approaches to producing discriminant functions and although the multiple linear regression method given here is conceptually the simplest, it is only applicable to situations where there are just two target categories. If the intention is to allocate individuals to multiple target categories, the reader is referred to canonical variates analysis (Chapter 10) and to more advanced texts on discriminant functions such as Digby and Kempton (1987) or Morrison (1967).

The example here concerns measurements made on oak leaves from two different species; red oak *Quercus rubra* and turkey oak *Q. cerris*. Ten leaves of each species were measured

Table 3.3 An illustration of the use of multiple linear regression to derive parameters.

This example uses multiple linear regression to solve the equation for the 'triangular function':

$$f(x) = \sum_{i=1}^{i=x} i$$

Hence $f(1) = 1$, $f(2) = 2 + 1 = 3$, $f(3) = 3 + 3 = 6$, $f(4) = 6 + 4 = 10$, etc. It will be assumed that the equation for this function takes the form:

$$F(x) = A + Bx + Cx^2 + Dx^3$$

Multiple linear regression can be used to find the parameters A, B, C and D, by setting up the following dataset:

Dependent variables $f(x)$	Independent variables		
	x	x^2	x^3
1	1	1	1
3	2	4	8
6	3	9	27
10	4	16	64
15	5	25	125

These data were entered into the standard package SPSS for Windows. The output file has been edited down to remove extraneous information.

R^2 1.00000

--------------- Variables in the Equation ---------------

Variable	B
x	.500000
x^2	.500000
x^3	4.82259E−15
(Constant)	−8.59868E−14

---

for five properties; length, breadth, petiole length, basal angle and number of lobes on the leaf. The preliminary results (Table 3.4) showed that the species differed greatly in the length of the leaves and of the petiole. Red oak leaves were generally larger than those of turkey oak (Figure 3.17), making it possible confidently to allocate a leaf to a species on the basis of these simple measurements. The dataset in Table 3.4 also includes two unknown individuals A and B, whose species may be predicted using the regression equations.

The approach to be taken here will involve calculating a linear equation which attempts to return a 1 for red oak and 2 for turkey oak. A standard multiple linear regression is set up with species number as the dependent variable, and biometric data as the explanatory variables. (Allocating a number to identify a category – here species – is an example of a **dummy variable**. In this case the dummy variable can take the value of 1 or 2, but any other pair of values would be equally usable.) The actual values of the regression coefficients are of little interest in themselves, but the overall regression equation defines a discriminant function which can be applied to other sets of measurements to provide a prediction of their species.

There are many possible regression equations that could be used here, depending on which biometric data are included. Table 3.5 gives some of the regression equations

Table 3.4 Oak leaf biometric data used to illustrate how multiple linear regression can generate discriminant functions. Abbreviations: sp – species (1 = red oak *Quercus rubra*, 2 = turkey oak *Q. cerris*); len – leaf length, mm; brth – maximum leaf breadth, mm; pet – petiole length, mm; ba – basal angle, degrees; nlobes – number of lobes on leaf.

sp	len	brth	pet	ba	nlobes
1	152	120	29	45	9
1	160	125	20	40	10
1	175	140	19	50	12
1	150	106	23	40	12
1	190	152	37	80	11
1	169	133	36	80	10
1	172	153	20	80	17
1	152	107	23	40	15
1	150	115	28	58	10
1	175	120	30	60	10
2	100	50	12	35	8
2	110	60	9	40	9
2	103	46	10	30	12
2	97	38	15	30	15
2	115	65	15	50	18
2	84	58	11	60	19
2	110	68	10	50	8
2	103	46	10	30	12
2	97	38	15	30	15
2	105	60	14	50	19

measurements on unidentified individuals

A: ?	105	57	10	40	11
B: ?	170	135	20	45	11

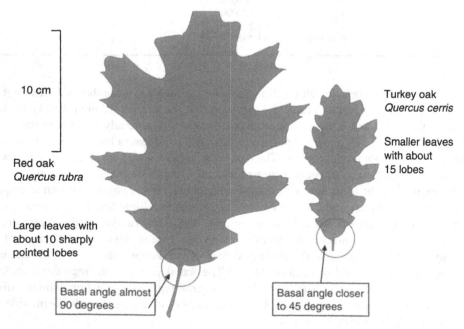

Figure 3.17 The leaves of turkey and red oaks, used to demonstrate the use of multiple regression to derive discriminant functions.

Table 3.5 A summary of the regression equations derived as discriminant functions to relate the oak leaf measurements to species (Table 3.4). Each line of the table gives the regression coefficients for the biometric variables under different models. The important point to make with this table is the difficulty in interpreting individual regression coefficients. Notice for example how the sign of the regression coefficient for leaf length changes from negative to positive when further variables are added. Despite this all these equations end up with high R^2 values making the same predictions, except for one analysis based on a random subset of the data which led to two misclassifications.

Source of data	Measurements used in model	Length	Breadth	Petiole	Basal angle	Lobe number	Constant	R^2	Number of misclassifications
All set	Length	−0.0143	–	–	–	–	3.412	0.890	0
All set	Breadth	–	−0.0119	–	–	–	2.571	0.883	0
All set	Length and breadth	−0.0082	−0.0053	–	–	–	3.066	0.900	0
All set	All measurements	0.0043	−0.0155	−0.0221	0.0133	−0.0064	2.182	0.970	0
Random half	All variables	0.0277	−0.0379	−0.0293	0.0273	0.0255	0.0321	0.991	0
Random half	All variables	−0.0020	−0.0093	−0.0263	0.0110	0.0037	2.5689	0.992	2

derived. The first (and simplest) example comes from simply using leaf length to predict species. In this case the regression equation is:

$$\text{Species} = 3.412 - 0.0143 \times \text{length}, \quad R^2 = 0.89$$

This equation predicts that unknown individual A has a species score of $(3.412 - 0.01413 \times 105) = 1.91$, while individual B has a score of 0.981. Rounding these to the nearest integer predicts that A is a red oak and B is a turkey oak. It is possible to check the accuracy of the function by comparing the predicted value of known individuals with their actual species: even for this simple equation the success rate is 100%. (This high success rate should not hide the fact that discriminant functions generally produce some false classifications. While a student at York University, my lecture group were asked to bring our height and weight to the lecture on discriminant functions, in order to predict our sex. My measurements predicted me to be the most female person in the class, despite being male! This was because the discriminant function used had been derived for a cohort of American students where the males generally weighed heavier than females, and my misclassification was due to a low body weight for my height.)

Adding more biometric variables to the oak leaf data increases the R^2 value but clearly cannot increase the success rate in this case. It is not important to worry about what the exact values of the regression coefficients mean, since they change greatly with minor changes in the model. The most important point to learn from Table 3.5 is that very different-looking sets of regression coefficients can predict nearly identical models. The data here are highly collinear, and the resulting sets of regression coefficients are unstable. What matters is the value that they predict.

It is generally recommended that a discriminant function should not be tested on the data used to derive it. One solution (known as jack-knifing) involves excluding one observation from data, deriving a discriminant function from the remaining data, then predicting the classification of the one excluded point using this new function. This process is repeated for the entire dataset, so that each observation has been classified using a function in which it played no part.

An alternative is to split a dataset into two randomly chosen subsets, derive a discriminant function for one half of the data, then test it on the other half. This has been done separately for two randomly chosen halves of the oak leaf data of Table 3.5, and the results are tabulated in the lower section of Table 3.5. Again the huge variation in regression coefficients should be noted: one half of the data generated the relationship:

$$\text{Species} = +0.0277 \times \text{length} - 0.0379 \times \text{breadth} \cdots + 0.0255 \times \text{lobes}$$
$$+ 0.0321 \text{ (with no misclassifications)}$$

while the other generated the function:

$$\text{Species} = -0.0020 \times \text{length} - 0.0093 \times \text{breadth} \cdots + 0.0037 \times \text{lobes}$$
$$+ 2.5689 \text{ (with two misclassifications)}$$

In fact it is a mistake to even try to interpret individual regression coefficients in collinear data such as these. This effect recurs whenever multiple linear regression is used, and is a common pitfall for researchers. It also intrudes into other techniques derived from multiple linear regression, such as the popular but complicated technique known as canonical correspondence analysis (Chapter 9).

3.6.4 Example: Using diatoms to reconstruct lake pH

Dixit (1986) studied diatoms (single-celled algae with a silica case) in lakes affected by the acidic gases from the Wawa smelter, Ontario. It is known that these algae can be identified from subfossil remains in lake sediments, and that different species of diatoms have different preferences for lake water acidity. These two factors mean that diatom assemblages can be used to reconstruct the acidity of lakes going back over many hundreds of years. The exact function used to relate diatom communities to pH is still contentious, and Dixit was concerned to explore four different models to see which gave the most reliable predictions of lake pH.

In order to achieve this, Dixit took diatom data from 28 lakes in Ontario whose pH values were known and ranged from 4.0 (strongly acidic) to 8.1 (basic). These pH values were then entered as the dependent variable in four linear regressions, to evaluate four competing models. Three of the models involved allocating diatom species to broad categories of pH preference, as follows: alkalibiontic – occurring only in pH values >7; alkaliphilous – occurring principally in pH values >7; circumneutral – occurring around pH 7 with an equal distribution above and below pH 7; acidophilous – occurring principally below pH 7; acidobiontic – occurring only below pH 7, and indifferent – showing no specific pH preferences.

These categories were used to calculate two indices; these indices followed the calculations of previous workers, and the weightings given below (3.5, 5, 40 and 108) are not derived from calculations based on Dixit's data. The reader is referred to Batterbee (1984) for further information on diatom indices.

$$\text{index } \alpha = \frac{5 \times \%\text{acidobiontic} + \%\text{acidophilous}}{5 \times \%\text{alkalibiontic} + \%\text{alkaliphilous}}$$

$$\text{index } \beta = \frac{\%\text{indifferent} + 40 \times \%\text{acidobiontic} + 5 \times \%\text{acidophilous}}{\%\text{indifferent} + 108 \times \%\text{alkalibiontic} + 3.5 \times \%\text{alkaliphilous}}$$

Indices α and β were found to need logarithmic transformation in order to normalise them prior to regression analysis.

The four competing models examined were:

1. pH is inferred solely from index α.
   Regression equation: pH $= A + B \times \log(\text{index } \alpha)$.
   $R^2 = 0.78$.
2. pH is inferred solely from index β.
   Regression equation: pH $= A + B \times \log(\text{index } \beta)$.
   $R^2 = 0.67$.
3. pH is inferred from the relative proportions of each group.
   Regression equation:
   pH $= A + B \times \%\text{alkalibiontic} + C \times \%\text{alkaliphilous}, \ldots$ etc.
   $R^2 = 0.89$.
4. pH is inferred from the actual percentage of species present.
   Regression equation:
   pH $= A + B \times \text{species 1} + C \times \text{species 2} + \cdots$, etc.
   $R^2 = 0.96$.

It was found that all four models gave highly significant correlations ($p < 0.001$), but that the R^2 values ranged from 67% (model 2) to 96% (model 4). It was concluded that the derived indices α and β were in fact rather poor predictors of pH, and that the best approach was to include all species where the community is known.

3.6.5 Example: Corncrakes and vegetation types

This final example is a more complex application of multiple linear regression than the datasets so far explored. At its core is a multiple regression analysis in which the dependent variable was the density of the endangered farmland bird the corncrake *Crex crex*, and the explanatory variables were environmental features. Due to the problem of pseudoreplication (Section 1.5.2) these multiple regressions had to be performed many times on different subsets of the data, and the final data analysis consisted of analysing the parameters generated by the entire set of multiple regressions. This procedure is known as jack-knifing (Efron 1979), and has been mentioned previously in connection with diversity indices (Section 2.4).

Green (1996) studied the distribution of the corncrake in relation to vegetation types in the UK. This is a bird of traditionally managed hay meadows, whose near extinction in Western Europe has been caused by intensification of agriculture. The study intended to discover what inferences about its habitat requirements could be inferred from the distribution of breeding pairs.

The basic theme of the study was to quantify the density of singing males (the only reliable way to survey these secretive birds) and vegetation types in the birds' strongholds of Scotland and Ireland, where traditional farming methods allowed their survival. Survey areas were divided into 1 km squares, in which dominant vegetation types and density of calling corncrakes were recorded. The basic design would therefore lend itself to a simple regression analysis:

$$\text{Corncrake density} = A + B \times \text{vegetation type 1} + C \times \text{vegetation type 2} + D \times \text{mowing date} \cdots$$

This model was indeed calculated, but there was an important complication relating to pseudoreplication which made the final analysis considerably more complicated than might at first appear necessary. The problem was that a key variable in the analysis was the date at which each meadow was cut: early cutting (as happens on modern farms) destroys the nest before the eggs have hatched. Date of mowing was therefore included as an explanatory variable, but it was not possible to obtain mowing dates for each meadow in the survey areas. (This would have involved interviewing a prohibitively large number of small farmers.) The solution adopted was to divide the data up into 16 geographical regions, within which the mowing date was assumed to be the same. The algorithm for deciding the mowing date for each region is given in the paper and need not be elaborated here: the crucial point is that only one estimate of mowing date was used for every site within a given region. This poses a serious limitation on subsequent analyses since different meadows within a given geographical region cannot now be handled as independent observations in any analysis that involves the estimated date of mowing. (As is normal in cases involving pseudoreplication, any computer package would perform the calculations without detecting any error or issuing any warnings – see Section 1.5.2 for more details.)

			A	B	C
Exclude region 1:	Corncrake $= A + B \times$ Iris $+ C \times$ Phragmites	$\longrightarrow$	−5.2	26.5	28.0
Exclude region 2:	Corncrake $= A + B \times$ Iris $+ C \times$ Phragmites	$\longrightarrow$	−6.2	25.9	27.1
....			.....		
Exclude region 16:	Corncrake $= A + B \times$ Iris $+ C \times$ Phragmites	$\longrightarrow$	−4.9	26.1	28.3
		Mean	−5.89	26.7	28.1
		S.e.	0.3	2.1	3.2
		$t = $ mean/s.e.	19.6	12.7	8.8

Figure 3.18 The basic idea behind jack-knifing to estimate regression parameters.

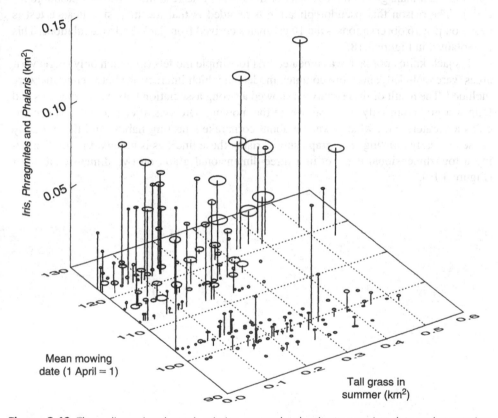

Figure 3.19 Three-dimensional graph relating corncrake density to mowing date and vegetation type. The width of the circles indicates corncrake number, ranging from 0 to the maximum of 10. (Reproduced by kind permission of the British Ecological Society.)

A simple, valid but crude approach would have been to calculate a mean value of vegetation cover and of corncrake density for each of the 16 geographical regions, then perform a simple multiple linear regression. Although valid, this would have lost most of the information in the dataset (which was at the scale of 1 km^2 and included a large amount of variation in plant cover and corncrake density) and would not have given reliable information about the small-scale factors affecting corncrake density.

Instead a more complex analysis was used based on the technique of jack-knifing. This involves repeatedly reanalysing the data after exclusion of one observation, then collating the results from these analyses and treating them as a dataset in its own right. This approach was implemented here by performing 16 separate multiple linear regressions, each using data at the finest-scale (1 km) resolution but excluding observations from one of the 16 geographical regions. This generated 16 separate estimates for each regression parameter. Each of these sets of 16 estimates is then treated as a new dataset, for which a mean and standard error were calculated. The value of the mean divided by its standard error gives the *t* statistic which tests the null hypothesis that the parameter tested actually takes the value zero. Essentially one is asking how many standard errors above zero the mean lies. Although clearly much more laborious than a straight-forward multiple linear regression, this approach manages to include data at the finest-scale resolution without pseudoreplicating. The reason that pseudoreplication is avoided is that the final significance test is based on just 16 observations – the 16 estimates derived from jack-knife calculations. This is explained in Figure 3.18.

This jack-knife approach was repeated first for simple models (in which only single variables were included) and more complex models in which interactions between terms were included. The result of these analyses showed a strong association between corncrakes and long wet grass, but only for areas where the mowing date was after mid-July. This coincides accurately with what is known about corncrake's nesting habits, and the mortality caused by early mowing. The graph summarising these findings is noteworthy for displaying a four-dimensional data set in a three-dimensional graph on two-dimensional paper (Figure 3.19).

Basic concepts in ordination

4.1 Introduction to ordination

4.1.1 What is ordination?

Most of the remaining chapters in this book concern different approaches to the ordination of data. The term ordination is rarely met outside of mathematics (the Chambers English Dictionary (Landau and Ramson, 1988) only defines ordination as 'admission to a Christian ministry'!) so needs careful introduction.

Ordination derives its name from the sense of putting information into order, or finding order within data. This is not the simple operation of sorting numbers into ascending/descending order, but the more abstract concept of finding a concise and useful summary of the patterns within multivariate data. Goodall (1954a) introduced the term and defined it as 'an arrangement of units in a uni- or multidimensional order', as opposed to 'a classification in which units are arranged in discrete classes'.

This book will follow Kent and Coker (1992) in distinguishing between two forms of ordination: direct and indirect ordinations. Direct ordinations are simple to describe and perform, involving the researcher organising observations along some known gradient. This may be the correct approach in cases where a single trend dominates the data, but it requires the user to know in advance what the major trend is. Indirect ordinations come in many forms, but all detect trends within data without the user needing to define end-points or gradients within their data. Indirect ordinations are powerful tools for probing and exploring multivariate data, and are now routinely used in many branches of environmental science. Throughout this book, the unqualified use of the word 'ordination' will signify an indirect ordination.

4.1.2 Direct ordinations

Direct ordinations involve describing the changes in a multivariate dataset along a pre-defined environmental gradient. This is the simplest approach to ordination and has a long history in many fields of environmental science. Examples include transect surveys of plant communities along altitude gradients, studies of community development through time (succession), and descriptions of soil conditions along catenas. It is worth examining a few examples of these because the conceptual model underlying direct ordination underpins the more sophisticated mathematical approaches which will be described later.

Direct ordinations which use time as the ordination axis describe the temporal development of a system – this usually involves community successions. One example from many is the work of Olsen (1958), who studied the succession on sand dunes at the southern end of Lake Michegan. The retreat of glaciers in this area has left a series of raised beaches, parallel to the shoreline but up to 17 m above its present level, whose vegetation shows several clearly distinct phases of colonisation. In the youngest areas (caused by blowouts of existing dunes exposing new, bare sand) the first colonist is a species of marram grass *Ammophila breviligulata*. This plant stabilises bare sand within a few years, but appears to need bare sand and typically disappears within twenty years. The second phase of dune stabilisation involves other sand grasses, along with fast-growing trees (*Populus deltoids*, along with *Prunus pumila* and *Salix* spp). Once these plants have colonised and stabilised the dune, the community is invaded by pines *Pinus* spp. Pine colonisation takes place 50–100 years after initial exposure of the sand. Between 100 and 150 years after the start of the succession, black oak *Quercus* invades, and forms dense stands which may be found on areas 12 000 years old, so appears to represent the local climax community. This succession is summarised in Figure 4.1, which shows the age of the site as the dominant axis. This kind of diagram is partly fictional, since there is no one location at which this chronosequence may be seen in its entirety, but is a useful tool for portraying and understanding the community changes.

Direct ordination may use gradients in space instead of time. Figure 4.2 shows the distribution of plant communities away from a stream edge in a floodplain forest (Hughes and Cass, 1997) although there are in fact no clear boundaries. A more complex example of the same approach is given in Figure 4.3, in which two environmental gradients are shown (elevation on the *X*-axis and moisture on the *Y*-axis), with the range of habitats associated with these conditions in the Sequoia National Forest (Vankat, 1982). The ordination space here provides a useful way of describing the conditions where each habitat type is found, and can readily be related back to actual field conditions.

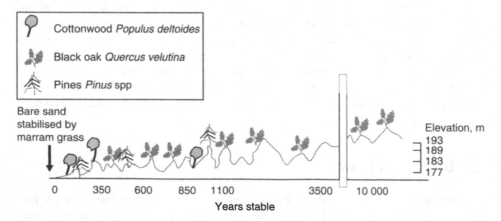

Figure 4.1 The sand dune succession at the southern end of Lake Michegan (redrawn from Olsen 1958). Reproduced by kind permission of Chicago University Press.

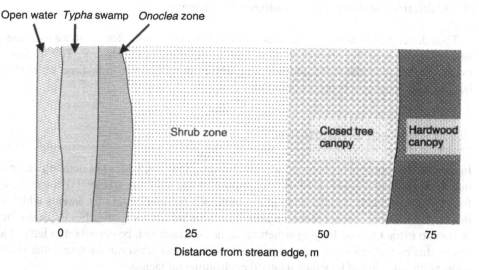

Open water *Typha* swamp *Onoclea* zone

Shrub zone

Closed tree canopy

Hardwood canopy

0   25   50   75

Distance from stream edge, m

Figure 4.2 Vegetation zonation away from a stream edge in a floodplain forest (redrawn from Hughes and Cass 1997 with kind permission of the British Ecological Society).

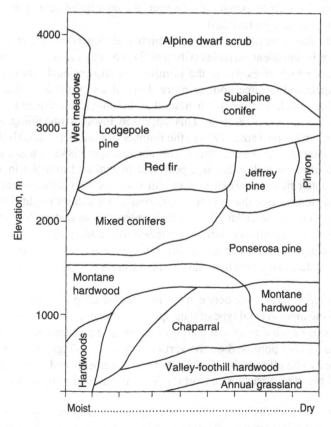

Figure 4.3 A mosaic diagram, in this case showing the distribution of vegetation types in relation to elevation and moisture in the Sequoia National Park. This is an example of a direct ordination, laying out communities in relation to two well-understood axes of variation. (Reproduced from Vankat (1982) with permission of the California Botanical Society.)

Thus direct ordinations are conceptually straightforward, but often involve a measure of subjectivity in that the researcher must define the gradient in advance. This limitation means that many textbooks on multivariate analyses do not even refer to direct ordinations, but move directly onto indirect ordinations.

4.1.3 Indirect ordinations

Indirect ordinations involve a more abstract form of analysis. Instead of defining an environmental gradient a priori based on observations in the field, the gradient(s) come(s) from the structure of the data itself. An indirect ordination produces a set of gradients which are inherent in the data, called ordination axes. There are numerous approaches to the problem of how to extract this underlying structure, some of which will be described in later chapters of this book. Only once ordination axes have been extracted can the researcher start to address the question of how they relate to environmental factors.

Thus indirect ordination can be seen as an approach to producing a meaningful summary of the patterns underlying multivariate data. This summary invariably takes the form of a picture or set of pictures (ordination diagrams) which allow the user to see relationships between data points. In order to explain this further, the concept of data space first introduced in Section 3.2 needs to be revisited.

We have already met linear regression, in which one dependent variable is plotted against one or more independent variables (Chapter 3). The most familiar example of this is the bivariate case, which gives rise to the familiar two-dimensional scattergraph. This can of course be extended to three, four or more dimensions by the inclusion of further independent variables. Each new variable is treated as defining a new axis in a new dimension, at 90 degrees to all previous axes. This approach involves imagining an abstract high-dimensional mathematical space (where the number of dimensions equals the number of variables being plotted against each other). This abstract space is known as a data space. Data space has no physical reality, but is a powerful mental tool to assist in visualising data. It does not matter that no normal human can visualise four-dimensional space, let alone the 20-dimensional space that a typical environmental dataset might define. Leave the calculations to a computer package! What is important is to realise that observations (i.e. lists of species from a quadrat or chemicals measured in a sample) become coordinates in a data space. Samples which consist of similar measurements will end up close to each other in data space; samples which have very different compositions will be widely separated.

There is one important difference between the model of data space used in multiple linear regression and the more general type of data space used in ordination. Linear regression involves labelling one variable as special and different from the others – this is the dependent variable, and is always portrayed on the vertical (y) axis on diagrams. The dependent variable is assumed to be affected by all the other measurements and is subject to measurement error. All the other variables in a regression are assumed to be measured without error (Figure 3.5).

Ordination makes no such distinctions between variables, and generally assumes that all are potentially of equal importance. (There are minor exceptions to this generalisation, such as options within some packages to downweight information about scarce species.) Thus the data space considered by ordinations is 'value-free', and will give the same results whichever order the information is presented. Figures 3.9 and 3.10 illustrate the difference between regression and ordination data space, showing how even for two-dimensional data, transposing the axes alters the regression line, while the best-fit line produced by an ordination technique (in this example principal components analysis) remains unchanged. For higher-dimensional data the regression plane varies greatly according to which variable is defined to be the dependent. Ordination is used in circumstances where there is no one clearly defined dependent variable, and instead the aim is to produce a useful condensed image of the entire dataset. Ordination techniques may safely be applied to data which have a high degree of co-correlation – indeed such datasets are ideal for the kind of data reduction offered by ordination.

Ordination techniques assist here because they reduce the dimensionality of data. Multivariate data is difficult to visualise because it contains too many dimensions to allow for easy plotting of all the possible graphs. A multivariate dataset can be imagined as an object in a high-dimensional space, which needs to be converted into a manageable number of dimensions (ideally two or three) to allow its structure to be seen. The problem is how to achieve this without causing undue loss of information or distortion of the shape.

A useful conceptual model of ordination techniques is that they take as input an object in high-dimensional data space, and produce as output an object in a new lower-dimensional data space. Both the old data space (the raw data) and the new data space (the ordinated data) consist of axes meeting at 90 degrees, but there are important differences between them.

The first point to consider is the meaning of the actual numbers on each axis. For the raw data this is easy – the axes correspond to measurements made by the researcher. For the ordinated data things are not so simple. The numbers on the axes are ordination scores. All ordination techniques involve calculating new variables, either called ordination or axis scores. For Bray–Curtis ordination these may be derived by hand, but for all other techniques a computer is required since the procedures used are long and complicated. Ordination scores are dimensionless (that is to say that they have no units) and look at first sight like random noise. When presenting an ordination to someone unfamiliar with the technique, it can be hard to answer the simple question 'what does this collection of numbers mean?'. It will be noted that the axis units on ordination diagrams bear no obvious relationship to the raw data (compare the axis scores for a PCA, say Figure 6.6, a CA, say Figure 7.1, and a DCA, say Figure 7.4). Indeed for some techniques (notably DCA ordination, the subject of Chapter 7), the sequence of operations between the raw data and the ordination diagram is so complex that an act of faith is needed to believe that the ordination describes the data at all! (In fact DCA has consistently proved to be one of the most useful of all techniques when tested on real data.)

The second difference between the raw data space and ordinated data space concerns the importance of each axis. For raw data all axes are equally important, and they can be put into the ordination in any order. This is not true for ordinated data. The axes produced by an ordination will be in descending order of importance, with the first axis being the most informative, the second axis the second most important, etc. Consequently the most useful ordination diagram is likely to be that which plots the first ordination axis against the second, although other combinations (first/third, second/third, etc.) will be worth checking. The reader will notice that I am being careful to avoid the word 'significant' here, because in statistical terminology significance is closely linked to calculations of probability. Ordination techniques do not directly provide probability levels, so that an ordination

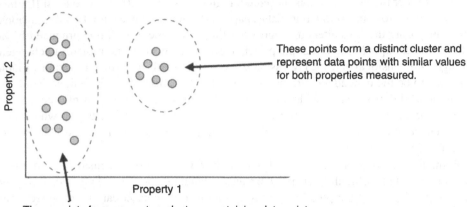

These points form a distinct cluster and represent data points with similar values for both properties measured.

These points form one or two clusters, containing data points with similar values for property 1.

Figure 4.4 Clustering of points in bivariate data.

cannot be said to be statistically significant. (However, the pattern produced by an ordination may be entered into further statistical tests which do give estimates of probabilities.)

4.1.4 The ordination diagram

There are many different indirect ordination techniques, of which the most commonly used are described in Chapters 6–9. In all cases the output will involve a set of diagrams which encapsulate information about the pattern of points within data space. The general principle is that points close together on an ordination diagram contain similar measurements, while points widely separated contain very different measurements. This can be seen with the bivariate data from Figure 4.4, which contains three clusters of points. Two clusters may be seen to be almost touching, while a third is well separated. In the case of bivariate data this result is trivial, since the two-dimensional graph may readily be constructed and inspected, but when dealing with multivariate data one can get swamped by the number of graphs to inspect.

Instead, for multivariate data, one constructs and inspects an ordination diagram. This will look superficially like a normal bivariate graph, except that the axes will be labelled not with the name of a measured variable but with an axis number (axis 1, axis 2, etc). These axes are the ordination axes, derived from the raw data by whichever technique has been invoked. Each observation can now be plotted as one point in this space (Figure 4.5), and points which lie close together can be assumed to be derived from similar observations.

There remains one important oddity to be understood about ordination diagrams. Although the order of their axes is precisely defined, the actual value of the numbers on their axes generally is not. Ordination scores may be multiplied by any arbitrary number (including negative ones) and still validly describe the shape of their raw data. (This is explained further in Section 6.3.6.) Consequently the same ordination diagram may validly be shown in four different ways A good analogy is to imagine that the diagram has been painted on a square sheet of glass. This sheet of glass can be rotated or flipped over, changing the actual pattern seen but not the underlying relationship between the points. (This is also true of direct ordinations: the mosaic diagram presented in Figure 4.3 could equally

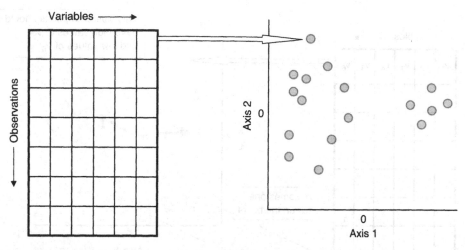

Figure 4.5 A basic ordination diagram looks and behaves just like a standard bivariate scattergraph. The difference is that each point plotted now defines the properties of the entire row of values collected for each observation, instead of just representing values of two variables.

be plotted with the X-axis running from dry to moist, and the Y-axis running from low elevation to high elevation.)

4.1.5 Biplots

Biplots are a useful extension of the ordination diagram, in which an ordination of properties of the columns of the matrix (the variables analysed) are overlain on top of the main ordination diagram in order to highlight relationships between the two sets of information. Such a diagram can rapidly allow the user to see which properties are strongly associated with each observation, since (as always on ordination diagrams) proximity implies close association.

There is a convention that this second set of information is visually distinguishable from the main ordination, usually by showing the values as arrows that run from (0, 0) to the coordinate in question. If such an arrow points clearly towards a cluster of points, it is expected that the variable in question takes particularly high values in the observations making up the cluster (Figure 4.6).

The actual methods used to set up the biplot differ according to the ordination technique used. For DCA (Chapter 7) and CCA (Chapter 9) ordinations the two sets of coordinates that are plotted are generated simultaneously and in the same manner, but a biplot generated by a principal components analysis (PCA, Chapter 6) contains two sets of information derived in rather different ways. In all cases the idea is to assist in the interpretation of an ordination by highlighting visually relationships within the data.

4.1.6 Setting up an ordination: The choice of variables

Given a dataset that might benefit from ordination, and a suitable statistical package, the procedure to follow is straightforward but (as is often the case with multivariate techniques)

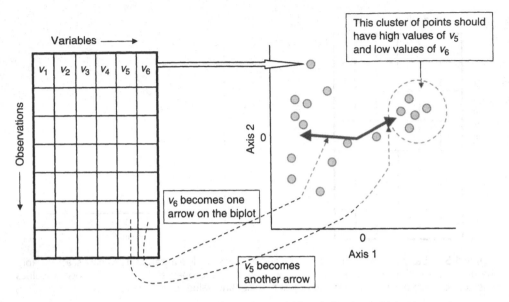

Figure 4.6 A biplot adds information about the variables, indicating which observations were distinguished by high levels of each marked variable.

contains an element of subjective choice. There is no division of the data into 'dependent' and 'independent' variables, unlike all inferential tests. The choices to make concern the selection of which variables to include in the ordination.

A reasonable first approximation is to say that one should include all variables that were measured and could reasonably be expected to contribute to a meaningful interpretation of the data. Thus for a study of soils one might include all numerical soil properties (pH, elemental contents, cation exchange capacity, etc.), for a morphometric study one might include all available body dimensions, for a study of a biological community one might reasonably include all species. If the ordination is to be purely descriptive, it would be quite legitimate to enter all available data. Each variable should be checked for normality and outliers, and it is standard practice to log-transform the data (which implies $\log(X + 1)$ for data containing zeros) prior to ordination – this will not harm the analyses and makes it more likely to pick up underlying trends.

It is however conventional to restrict an ordination to just one type of variable. Typically an ordination of (for example) a biological community would include most or all of the species, but exclude physical/chemical data. An ordination of a set of soil chemical properties would normally exclude data on climatic data or biological characteristics. The reason for this is to allow statistically valid analyses of the relationship between the resulting ordination scores and environmental variables. (This is explained in more detail in Section 6.5.) One important exception to this principle comes with the use of canonical techniques such as CCA ordination (Chapter 9), which specifically requires that two blocks of different types of data are ordinated together.

There then comes the more subjective decision about the exclusion of variables that are valid but appear to be of little importance. This is a particular problem with biological data where most species are uncommon, containing mainly zeros. Intuitively one can appreciate that inclusion in a large dataset of a plant that occurred as a single individual in one observation is unlikely to be very useful – but where to draw the line? One of the differences in ordination techniques lies in their sensitivity to the inclusion of scarce species; Bray–Curtis ordination (Chapter 5) and principal components analysis (Chapter 6) are relatively robust, while correspondence analysis and DCA (Chapter 7) are prone to over-emphasising the importance of rare species. The best approach is probably to repeat the ordination several times with differing species lists (all species, all excluding rare species, then using common species only) to examine the stability of the conclusions. To have faith in the usefulness of an ordination, its interpretation should be robust over a range of species exclusion thresholds.

4.2 A historical overview of ordination

Until the appearance of Bray–Curtis ordination (otherwise known as polar ordination) in 1957 (Bray and Curtis, 1957), ecologists had generally described communities (whether animal or vegetable) using simple scatterplots to describe variation in individual species rather than producing indices that covered the entire community. Bray–Curtis ordination is described in Chapter 5, and can be summarised as being a procedure in which the researcher identifies two opposing end-points in their data (the start and end of a time series, or extremes of an environmental gradient such as elevation above sea level). The procedure then uses simple calculations to order all the other observations along a line running between these two extremes. It entered the research environment more for its convenience of use (this was in the largely precomputer age) than its ability to reveal insights into data.

The era of easy access to digital computers started in the 1960s, when universities started to own mainframe computers. These mainly ran programs written in the language FOR-TRAN, but were able to perform the huge iterative calculations needed for indirect multivariate ordinations. The ordination of choice in this period was principal components analysis (PCA, also frequently called factor analysis, which is covered in Chapter 6), which has the virtue of (relative) mathematical simplicity. This ordination method was described as far back as 1933 (by Hotelling), and was applied to ecological data by Goodall (1954b) under the name of factor analysis, but use of PCA only became widespread after it was described by Orloci (1966). A second ordination whose use by environmental scientists (especially ecologists) became widespread towards the end of the 1960s was correspondence analysis (CA, alternatively known as reciprocal averaging, which is covered in Chapter 7). This technique has in fact been reinvented independently many times (De Leuw, 1983), and has the advantage that observations are ordinated in the same operations as variables.

There is one ordination technique which dates to the early 1970s but has been increasing in popularity throughout the 1990s, namely non-metric multidimensional scaling or NMDS (covered in Chapter 10). This was introduced by Anderson (1971), and has the drawback of being highly demanding of computer time.

In 1973 Mark Hill introduced two new packages which led to a quiet revolution in the analysis of environmental data. The first was a program known as DECORANA (covered in Chapter 7), which was a modification of correspondence analysis intended to iron out some of the wrinkles known to be associated with this ordination. The algorithm implemented by DECORANA is properly known as detrended correspondence analysis (DCA). The actual operations involved in this ordination are complex and involve repeatedly reshaping the data, leading to persistent queries about the reliability of the technique. Despite this it has stood the test of time remarkably well, so that almost 20 years later Kent and Coker (1992) stated of the search for a 'best' ordination method:

'where [environmental] data are not available, DCA probably remains the best and most appropriate choice'.

The second release was a program called TWINSPAN, for the classification of ecological data, based on dichotomously dividing observations based on a correspondence analysis of the data. This program is described in Chapter 8, and it remains a standard tool for numerical classification at least within the ecological community.

The most recent ordination technique to become widely used by the environmental research community is canonical correspondence analysis (CCA, described by ter Braak in 1986). This is another derivation of correspondence analysis, but now the primary data (usually species) are constrained to optimise a fit to a secondary matrix (usually environmental data). This has become the ordination of choice by researchers wishing to summarise relationships between two independent multivariate datasets, such as between species composition and environmental factors.

A recent but promising development allows users to view the impact of experimental treatments on a multivariate object (such as an ecological community). This is known as principal response curves (PRC), and is described in Chapter 10.

5

Bray–Curtis ordination

Bray–Curtis ordination (otherwise known as polar ordination) was the earliest indirect ordination technique to be widely adopted by ecologists following its introduction in 1957 (Bray and Curtis, 1957). This was largely due to the simplicity of the actual algorithm, which can readily be performed by hand – a big advantage in the days when computers were large, expensive and slow. (Even so, the amount of work involved in performing a Bray–Curtis ordination of a typical environmental dataset by hand is substantial.)

The technique was originally used to analyse data from the forests of Wisconsin, USA, where vegetational gradients existed which reflected a complex of environmental and biotic interactions. One over-arching force shaping the vegetation structure was elevation, and one approach to analysing these data would have been to plot the frequency of occurrence of each species as a function of altitude. This would have been an example of direct ordination. Instead of using direct ordination to show altitudinal variation, Bray and Curtis devised an objective technique to graph plant communities on axes which reflected actual differences in vegetational composition. This technique was popular for a few years until superseded by eigenvector-based ordination techniques, and is hardly ever used in modern research papers. A search on a major scientific database found just two research papers which used this ordination technique in the period 1990–1996. This decline is reflected by the fact that only one modern multivariate package offers Bray–Curtis ordination as a standard feature (MjM Software Design, 1995). This fall from grace results from a perception that Bray–Curtis ordination is an 'old fashioned' technique, which is only partially true. Beals (1984) has defended this ordination technique as a useful and effective alternative to standard ordinations. Its use appears to have been confined to ecological data, although in principle it could be extended to a wide variety of other types of data providing a suitable similarity index is chosen.

The technique is introduced here because it is a conceptually easy introduction to indirect ordination techniques which can be performed by hand, using nothing more sophisticated than a pocket calculator, ruler and compass. In addition, the procedure followed has many similarities with the much more commonly used technique of cluster analysis, which will be covered in Chapter 8.

5.2 Theory of Bray–Curtis ordination

The basic procedure of Bray–Curtis ordination is shown in Figure 5.1. The starting point is a rectangular matrix of data describing the attributes of a set of samples (or observations). Usually these are records of the composition of a biological community, although other types of data could equally be used. This is converted into a square matrix that quantifies the differences (dissimilarities) between each pair of samples. Two samples are then chosen to represent extreme ends of the first ordination axis (the dominant underlying trend within the data). These are used to construct a scale diagram which defines the positions of all remaining samples on this ordination axis. A second pair of observations may be chosen to define a second ordination axis, and the process repeated.

5.2.1 The dissimilarity matrix

This ordination technique relies on the construction of a dissimilarity matrix. This is a matrix in which each element is a number which represents the difference between the communities in two different samples. Such numbers are known as dissimilarity indices, and have the property that zero values show two samples to be identical, while progressively larger values show increasing differences between samples. The choice of dissimilarity index used is important, since different indices will result in different ordination diagrams (sometimes

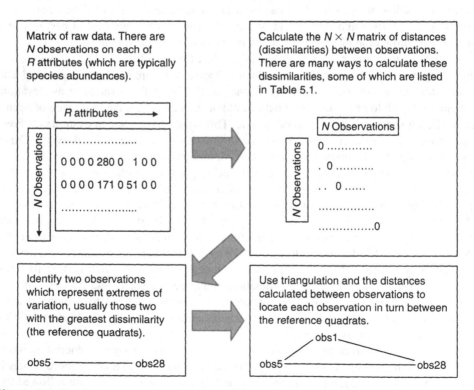

Figure 5.1 The three steps in a Bray–Curtis ordination.

dramatically so). Most such indices were constructed with biological datasets in mind, and describe the frequency with which species co-occur. These indices would not be suitable for datasets where all entries are non-zero (such as chemical properties of a set of samples, scores from questionnaire returns, etc). However one of the indices described in Table 5.1, the Euclidean distance between samples, would be applicable to non-biological data.

The index most commonly used in Bray–Curtis ordination is the Sørensen index, also known as the Bray–Curtis index or the Czekanowski index (Czekanowski, 1913). This and some other measures of dissimilarity are defined in Table 5.1. Table 5.2 gives worked examples of the calculation of these indices.

Table 5.1 Measures of dissimilarity often used for Bray–Curtis ordination. In each case the formula gives the distance function between two observations I and H (each observation consisting of a set of P species values $S_1, S_2, ..., S_P$), written as D_{ih}.

Name	How calculated	Comment
Czekanowski distance, also known as the Bray–Curtis distance, and the city-block metric	$D_{ih} = 1 - \dfrac{2 \times \sum_{j=1}^{j=p} \mathrm{Min}(I_j, H_j)}{\sum I + \sum H}$ This is also written as $D_{ih} = \dfrac{\sum_{j=1}^{j=p} |I_j - H_j|}{\sum I + \sum H}$ where $|x|$ means that the sign of the number is ignored (hence negative values are treated as if they were positive).	This is the default choice for a distance measure in Bray–Curtis ordination, more for its empirical effectiveness than any theoretical reasons. It is city-block metric closely related to the city-block distance or Manhattan distance (Table 8.4).
Relative Czekanowski distance	All observations are converted to a proportion of the total for their sample unit (recoding species by proportion or p_i values as shown in Table 2.1) then the distance calculated as before.	Faith *et al.* (1987) called this method the relativised Manhattan distance, and found it to be an under-used approach.
Euclidean distance	$D_{ih} = \left(\sum_{j=1}^{j=p} (I_j - H_j)^2 \right)^{0.5}$	This is simply the extension into multidimensional data space of the familiar distance we measure with rulers and tapes. It is not as suitable for ecological data as the Bray–Curtis distance as it is sensitive to outliers.
Jaccard distance	$D_j = b + c/(a + b + c)$ where: D_j = the Jaccard distance between samples 1 and 2, a = number of species common to samples 1 and 2, b = number of species only found in sample 1, c = number of species only found in sample 2.	This is derived as $1 -$ the Jaccard similarity index between two samples: $S_j = \dfrac{a}{a + b + c}$ where S_j is Jaccard's similarity coefficient.
Sørensen distance	$S_s = b + c/(2a + b + c)$ using the notation of the Jaccard index.	Beals (1984) found that this method was less satisfactory than the Sørensen distance. It only needs presence–absence data. Similar to the Jaccard index but is preferred because it concentrates on species that are common to both samples. It only needs presence–absence data.

Table 5.2 A worked example of calculating dissimilarities between two observations, using the first two lines of the Liphook forest fungal data set (Table 1.9). The table starts with the two lines of data (counts of individual species). For each pair of counts the minimum value is tabulated, then the three rows (two of data and one of minima) are summed. The Sørensen distance between the two observations is then calculated from the formula given in Table 5.1 and is found to be 0.316 or 31.6%.

	Boletus subtomentosus	*Cortinarius semisanguineus*	*Gomphidius roseus*	*Inocybe lacera*	*Laccaria laccata*	*Lactarius rufus*	*Paxillus involutus*	*Suillus bovinus*	*Suillus luteus*	*Suillus variegatus*	
Year: 86　Plot: 1	0	0	0	0	280	0	1	0	0	0	Sum: 281
Year: 86　Plot: 2	0	0	0	0	171	0	51	0	0	0	222
											0
minimum value	0	0	0	0	171	0	1	0	0	0	172

Bray–Curtis distance = $1 - (2 \times 172/(281 + 222)) = 0.3161$

The relative Sørensen distance is calculated the same way, after converting each value to a proportion of the total for its observation.

	Boletus subtomentosus	*Cortinarius semisanguineus*	*Gomphidius roseus*	*Inocybe lacera*	*Laccaria laccata*	*Lactarius rufus*	*Paxillus involutus*	*Suillus bovinus*	*Suillus luteus*	*Suillus variegatus*	
Year: 86　Plot: 1	0	0	0	0	0.9964	0	0.0036	0	0	0	Sum: 1
Year: 86　Plot: 2	0	0	0	0	0.7703	0	0.2297	0	0	0	1
minimum value	0	0	0	0	0.7703	0	0.0036	0	0	0	0.7739

Relative Sørensen distance = $1 - (2 \times 0.77/(1 + 1)) = 0.2262$

Euclidean distances are calculated by summing squared differences between values:

	Boletus subtomentosus	*Cortinarius semisanguineus*	*Gomphidius roseus*	*Inocybe lacera*	*Laccaria laccata*	*Lactarius rufus*	*Paxillus involutus*	*Suillus bovinus*	*Suillus luteus*	*Suillus variegatus*	
Year: 86　Plot: 1	0	0	0	0	280	0	1	0	0	0	
Year: 86　Plot: 2	0	0	0	0	171	0	51	0	0	0	
difference between values	0	0	0	0	109	0	−50	0	0	0	
difference between values squared	0	0	0	0	11 881	0	2500	0	0	0	Sum: 14 381

Euclidean distance = $\sqrt{14\,381} = 119.92$

5.2.2　Defining the end-points

Whatever dissimilarity index is used, a dissimilarity matrix is then constructed showing the dissimilarity between each pair of samples. A dataset with N observations will generate an $N \times N$ matrix, so that the work goes up sharply with increasing size of dataset. (Table 5.3 shows part of the 35×35 matrix of Sørensen distances between observations in the Liphook pine forest fungal data set, Table 1.9.)

It will be noted that the leading diagonal (top left–bottom right) of the matrix consists wholly of zeros. This simply means that there is no dissimilarity between a sample and itself. Like all dissimilarity matrices, this matrix is symmetrical about its leading diagonal;

Table 5.3 Part of the matrix of Sørensen distances between observations in the Liphook pine forest fungal data. (The full matrix is 35 × 35 so does not readily fit on the page.)

OBS: 1	2	3	4	5	6	7	8	9	10	11	12	13	14	15	16	17	18
1	0.00																	
2	0.32	0.00																
3	0.92	0.69	0.00															
4	0.75	0.68	0.66	0.00														
5	0.98	0.91	0.58	0.85	0.00													
6	0.78	0.69	0.54	0.13	0.65	0.00												
7	0.82	0.66	0.31	0.32	0.61	0.20	0.00											
8	0.57	0.73	0.98	0.92	0.99	0.93	0.95	0.00										
9	0.39	0.54	0.89	0.88	0.97	0.88	0.87	0.34	0.00									
10	0.40	0.18	0.69	0.68	0.91	0.69	0.66	0.75	0.49	0.00								
11	0.38	0.58	0.95	0.88	0.98	0.88	0.90	0.46	0.20	0.53	0.00							
12	0.52	0.25	0.53	0.52	0.85	0.53	0.49	0.83	0.67	0.27	0.73	0.00						
13	0.04	0.28	0.90	0.73	0.96	0.75	0.79	0.60	0.41	0.37	0.41	0.48	0.00					
14	0.15	0.23	0.76	0.75	0.93	0.76	0.73	0.62	0.37	0.22	0.41	0.41	0.12	0.00				
15	0.72	0.82	0.99	0.95	1.00	0.96	0.97	0.25	0.49	0.80	0.52	0.89	0.73	0.73	0.00			
16	0.50	0.67	0.97	0.90	0.99	0.92	0.93	0.52	0.31	0.63	0.19	0.79	0.52	0.53	0.40	0.00		
17	0.53	0.62	0.97	0.89	0.99	0.91	0.92	0.73	0.56	0.58	0.37	0.76	0.52	0.49	0.55	0.20	0.00	
18	0.77	0.86	0.99	0.96	**1.00**	0.97	0.97	0.45	0.57	0.84	0.56	0.91	0.79	0.79	0.38	0.45	0.52	0.00

The largest value in this entire matrix is between observations 5 and 18 (shown in bold as 1.0, although its exact value is 0.998182), so these are chosen as reference quadrats.

this means that the dissimilarity between sample A and sample B is the same as that between sample B and sample A.

Once the dissimilarities between all of the samples have been found, two samples need to be identified which represent the end-points of the ordination axis. In vegetation analysis these two samples are generally known as the **reference quadrats**. There are several methods for deciding which samples to use. The simplest is known as **direct end-point selection**, in which the user specifies two samples which represent perceived extremes of a known gradient. This makes subsequent interpretation simple, since the meaning of the ordination axis is defined in advance, but clearly is open to a charge of being subjective. Probably the best general solution is to define as end-points the two quadrats with the greatest dissimilarity (as listed in the distance matrix). In the case of the Liphook toadstool succession the greatest dissimilarity came between observations 5 and 18 (plot 5 in 1986 and plot 4 in 1988, between which the Sørensen distance was 99.8%).

5.2.3 Obtaining ordination scores

However the reference quadrats are identified, the exact positions of the other samples in this ordination space can now be defined by drawing the sample dissimilarities on a simple scale diagram. The first ordination axis is drawn on a sheet of paper, with the reference quadrat samples marked at its ends. This distance between these two samples needs to scale to the actual value of the dissimilarity between the samples. Thus if the dissimilarity between the reference quadrats were 1.0 or 100% (no similarity at all), the axis could be made 10 cm long, defining a scale of 1 mm = 1% difference.

Now the positions of the remaining samples is found by their distances from each end of the ordination axis. Thus (using a scale of 1% = 1 mm) a sample which is 40% different

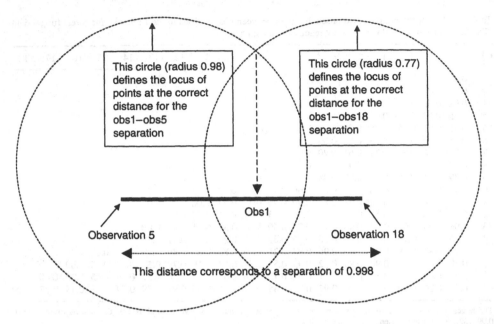

Figure 5.2 Converting Bray–Curtis distances into separation in an ordination space: an example using the distances presented in Table 5.3. The reference observations are numbers 5 and 18. To ordinate observation number 1 between these two points we need the Bray–Curtis distance between observations 1 and 5, and also the distance between observations 1 and 18. These values are 0.98 and 0.77 respectively. A circle is then drawn from observation 5, whose radius corresponds to 0.98 (the distance between observations 5 and 1), and a second from observation 18 whose radius corresponds to 0.77. Where the two circles intersect is the first axis position for observation 1.

from one reference quadrat and 70% different from the other, lies at a point 4 cm from one end and 7 cm from the other. This can readily be found by drawing circles of the appropriate radius from each end, which should intersect at two points (Figure 5.2). The deviations of these points off the ordination axis is ignored, but their projection onto the axis is their ordination score (Figure 5.2).

It will be evident that the ordination scores may be derived without construction of a scale diagram, by use of Pythagoras' theorem. An example is shown in Figure 5.3, and this calculation forms the basis of software implementations of Bray–Curtis ordination.

5.2.4 Second and higher ordination axes

A second ordination axis may be fitted in a similar manner, giving a two-dimensional ordination diagram. As for the first axis, the main problem lies in defining which samples to use as the end-points of the ordination (the second axis reference quadrats). The original approach was for the user to designate two samples which lay near the middle of the first ordination axis but which appeared to have widely divergent communities. This procedure can be criticised on several grounds. The selection of end-points for the second axis is clearly subjective, and the resulting axes are not orthogonal (they do not lie at 90 degrees to each other).

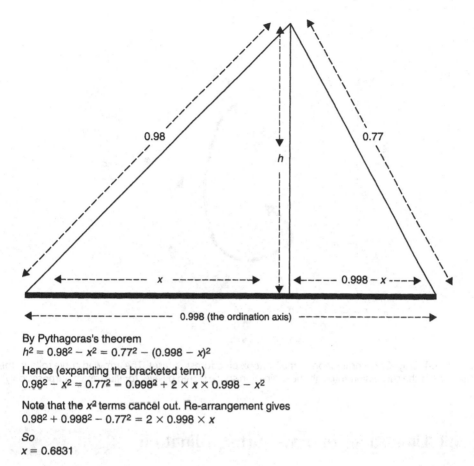

By Pythagoras's theorem
$$h^2 = 0.98^2 - x^2 = 0.77^2 - (0.998 - x)^2$$

Hence (expanding the bracketed term)
$$0.98^2 - x^2 = 0.77^2 - 0.998^2 + 2 \times x \times 0.998 - x^2$$

Note that the x^2 terms cancel out. Re-arrangement gives
$$0.98^2 + 0.998^2 - 0.77^2 = 2 \times 0.998 \times x$$

So
$$x = 0.6831$$

Figure 5.3 The Bray–Curtis ordination of Figure 5.2, re-expressed as a geometrical problem. The question is to find the distance X given the other measurements. The height of the triangle h is needed for the solution, but does not feature in the ordination output.

MjM Software Design (1995) give an improved method for the fitting of second and higher axes to Bray–Curtis ordinations, in which the original dissimilarity matrix is replaced by a matrix of residual distances which excludes the effect of the first axis. This allows an objective choice of end-points and ensures that subsequent axes are orthogonal.

However the end-points are defined, the same procedure of axis construction/scaled distance fitting using the original dissimilarity matrix is then followed, to give a second position for each sample. This means that each sample has a position on two ordination axes, and may be plotted on a two-dimensional ordination diagram.

The full Bray–Curtis ordination of the Liphook fungal dataset is given in Figure 5.4, showing a general separation on the first axis between the first two years and later, although the succession has been distorted into a U shape. This bending will prove to be a recurring feature of ordinations on this dataset, and will be returned to (with a solution) in Chapter 7.

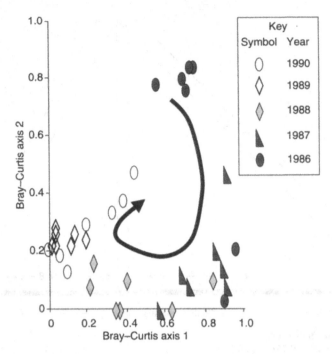

Figure 5.4 Bray–Curtis ordination of the Liphook pine forest dataset. The arrowed curve follows the trajectory of the succession through the ordination space (fitted by eye, for illustrative purposes only).

5.3 Limitations of Bray–Curtis ordination

The explanation given above of this ordination technique has highlighted some potentially problematic features of Bray–Curtis ordination. The technique has been criticised as being subjective, in that arbitrary selection of end-points is a recognised version of the technique. This is not necessarily a problem, since there is a choice of objective alternatives. The range of choice is in fact a problem in itself, since there are four different standard options for the dissimilarity index to be used, five methods of end-point selection for the first axis, and two methods of end-point selection for the second axis. This gives 40 different permutations, each a different technique and capable of giving different answers. (This problem of a plethora of permutations will be met again in greater force in cluster analysis, Chapter 8.) Different choices can often give very different-looking ordinations. Sometimes the maximum dissimilarity is shared by several different pairs of samples, leaving the user with a subjective choice of which pair to use as the end-points.

Secondly the standard implementation of algorithm (basing end-points on maximum values of the Czekanowski dissimilarity index) is very sensitive to outliers, which by definition will tend to be very dissimilar to other samples so will tend to be picked as end-points. The trouble is that one odd data point is unlikely to be a useful reference to fit all other samples onto an ordination axis.

Thirdly the geometry may simply fail to work. It is possible when using the Czekanowski similarity index to find situations in which the two overlapping circles shown in Figure 5.2

actually fail to touch. This can arise because this similarity index is not a true metric coefficient of similarity, so that the pattern of points in data space becomes distorted in projection to a lower-dimensional space. If this problem occurs, the only solution is to re-try the ordination another way, usually by trying a new pair of end-points.

As a final point, although Bray–Curtis ordination can be performed by hand, it should not be thought that this is a quick or effortless process. The amount of work goes up with the square of the number of samples, and (like all ordinations) is best left to computers for all but the smallest datasets.

6

Principal components analysis

6.1 A brief overview of principal components analysis

For the sake of convenience the multivariate technique known as principal components analysis will hereafter be referred to by its acronym PCA. Like all ordination techniques, PCA is a tool to condense data. PCA is the simplest ordination technique which is wholly objective (in contrast to Bray–Curtis ordination, which requires users to make some subjective choices). In standard English usage one could say that PCA is useful for situations where a set of data has been shaped by a few underlying factors. The factors may be interspecies differences underlying a set of body size measurements, soil differences underlying a set of plant community data, geological differences underlying a set of geomorphological data, or social differences underlying questionnaire returns. Intuitively one can appreciate that one such important underlying factor can have a profound effect on a great many aspects of a set of measurements. To take just one example, a change in geology between two areas can manifest itself as differences in soil, vegetation, valley depth, stream chemistry, land usage, and many other aspects of the environment. Any one of these features could be measured and turned into a data matrix in which the underlying pattern was the contrast between two rock types.

Unfortunately the term 'factors' should be avoided when explaining PCA, since it has been usurped by a related multivariate technique, factor analysis (FA). FA in the strict sense of the term is harder to interpret than PCA and much less widely used. FA assumes that there exist a small number of invisible (unobservable) underlying factors shaping your data, then tries to estimate numeric values for these unobservable factors. Factor analysis is given a brief further description in Chapter 10. Oddly, current practice in the USA is to use the term 'factor analysis' to refer to PCA. At least one major statistical package at the time of writing does supply PCA under the menu title of factor analysis, and produces an output that is a blend of PCA and factor analysis (SPSS, 2000). Such terminological confusion merely serves to complicate a field that many already find quite complicated enough. Another catch to watch out for is a multivariate technique related to PCA known as principal coordinates analysis (PCO), briefly described in Chapter 10. The two techniques should not be confused, but their names are so close that Chatfield and Collins (1980) suggest renaming principal coordinates analysis as classical scaling to prevent misunderstandings (a suggestion that has not been widely taken up).

The essence of PCA involves treating a set of data as a geometrical object, then finding the most informative angle from which to view the shape. The mathematics underlying PCA were first derived by Peerson (1901), and rediscovered by Hotelling (1933). It became widely adopted by plant ecologists in the 1950s, and remains the ordination technique of first choice for the majority of environmental scientists. PCA is the default ordination technique on most packages which perform ordinations, due to its ease of calculation, and robustness to deviations from multivariate normality in data.

In view of this suite of positive features, it may seem curious that PCA is not at present routinely included in undergraduate statistics courses. This is partly because of the perception that multivariate techniques are too complicated for non-mathematicians, and partly because there are genuine limitations to PCA (in common with all ordination techniques).

To appreciate these limitations, it is important to underline what PCA is *not*. It is not a test of significance, nor is any null hypothesis required. There is no objective procedure for deciding which variables to include prior to the ordination, nor to decide which were unimportant after the analysis has been run. The technique cannot cope with missing values, and may be inappropriate in situations where communities develop steadily along environmental or temporal gradients (such cases are better handled by DCA ordination due to a distortion called the arch effect, described in Chapter 7, Section 7.3).

PCA is not an end in itself, but should be followed by further lines of investigation – as a minimum a visual examination of the ordination diagram should be made. It is good practice to follow this up by generating a hypothesis about what the ordination is showing, then testing this by the application of a standard inferential test.

6.2 What does PCA achieve?

6.2.1 An analogy with shadows

A useful explanation would be to start by considering a dataset consisting of N variables measured together, such as the percentage cover of plant species in a series of quadrats along a transect. It is a standard – and good – piece of statistical advice to start any analysis by looking at your data. A good start to interpreting the data would be to plot all the data together. In the case of a dataset with just two independent variables this would produce the familiar two-dimensional scattergram. The concept of data space was introduced in Chapter 3 (Figures 3.1 and 3.2); plotting two variables against each other involves creating a two-dimensional data space. Four variables would require a four-dimensional data space (Figure 6.1), while with ten variables the scattergram would be in ten-dimensional space! No normal human can visualise this space, nor are there techniques available to produce a physical representation of such a high-dimensional graph. It would be possible to plot each species against each other using standard bivariate graphs, but for just ten variables this generates 45 graphs and the results would be extremely difficult to digest.

The intention of using PCA is to simplify data by reducing the dimension of the data space. The approach used by PCA can be explained in a conceptually simple way: it casts a shadow from the high-dimensional data space onto a low-dimensional space – typically a 2D scatterplot – where humans can visualise its patterns. Figure 6.2 shows how progressively higher-dimensional shapes may be visualised using a two-dimensional projection.

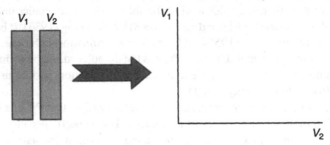

Two variables define a two-dimensional data space

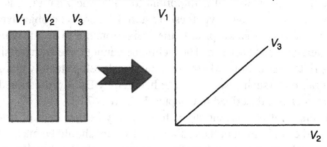

Three variables define a three-dimensional data space

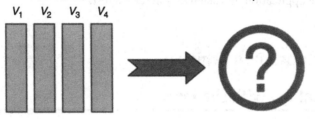

Four variables define a four-dimensional data space

Figure 6.1 The number of variables to be analysed determines the number of dimensions in the data space.

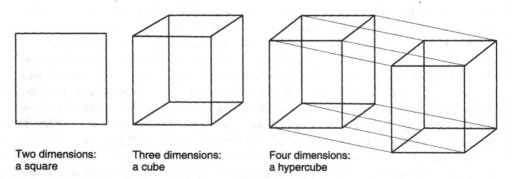

Two dimensions:
a square

Three dimensions:
a cube

Four dimensions:
a hypercube

Figure 6.2 An example of how a two-dimensional shadow image may be used to aid in the visualisation of high-dimensional shapes. Here the 3D and 4D versions of a square (a cube and hypercube respectively) are shown as undistorted shadows.

The shape to be visualised is a simple square and its higher dimensional analogues (the cube and the hypercube). A square is readily presented on paper, but a cube (being three-dimensional) needs some form of projection. The projection used in Figure 6.2 is a shadow, and the resulting pattern shows clearly enough how a cube consists of two squares with all the corners (vertices) attached. There are penalties inherent in this projection since it discards information about one dimension. One aspect lost in this projection is that it is no longer possible to see that all 12 sides are of equal length. Another drawback is that there are infinitely many different ways of casting the shadow (depending on the relative positions of light source, object and screen).

This approach can be extended to objects in higher dimensions, which humans cannot visualise. Figure 6.2 shows one possible shadow that a hypercube (the four-dimensional analogue of a cube) could cast in a two-dimensional space. Although we will never encounter such an object, the shadow allows us to see that it consists of two cubes with all the vertices attached. This simple example shows how 2D approximations can help humans understand high-dimensional shapes. PCA operates on datasets in very much this way, casting a shadow of a high-dimensionality object to make it accessible to humans.

It must be emphasised that the imagery of shadows used here is an unconventional approach to explaining PCA. No books deriving the algebra that underlies PCA will refer to shadows, as the mathematics are not strictly identical. The analogy is used here because it provides an easy way to visualise the procedure, but it should not be taken too far. In particular, light rays emerge from a point source in the shape of a cone, but the type of projection used in PCA would involve strictly parallel light beams. The shadow cast by a very distant light source such as the Sun would be an acceptable approximation.

The analogy with a shadow reveals an immediate problem. Where should the light source be located? The choice of where to locate the light source allows an infinite variety of possible shadows.

PCA can be thought of as starting off by finding the most useful, informative viewpoint from which to visualise the cloud of data points. It then casts a shadow giving an undistorted image from this viewpoint. Further possible shadows are cast, each using viewpoints that are at 90 degrees to all previous viewpoints.

Thus, in this interpretation, a PCA generates three sets of information from a dataset. The obvious result is that it produces low-dimensional images of high-dimensional shapes. Less obviously, it gives information about the position of the light source, hence the most productive angles from which to view the data. The third set of information that can be derived is the 'usefulness' of each viewpoint, which as stated sounds vague but in fact can be defined precisely in terms of a percentage of overall variance.

6.2.2 An alternative explanation

Having used the unconventional but accessible analogy with shadows to introduce PCA, it is essential to give a more conventional approach towards explaining the technique. As before, the explanation starts by considering each observation as a coordinate in N-dimensional data space, where N is the number of variables and each axis of data space is one variable (Figure 6.1).

Now a new set of coordinate axes are drawn on the graph, whose origin is at the overall mean of the data (Figure 6.3).

Step 1: A new set of axes is created, whose origin (0,0) is located at the mean of the dataset. These new axes are shown as dashes.

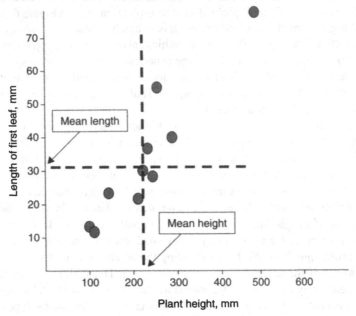

Step 2: The new axes are rotated about their origin until the first axis gives a least squares best fit to the data. Notice that the residuals (arrowed) are fitted orthogonally not vertically (contrast with Figure 3.5).

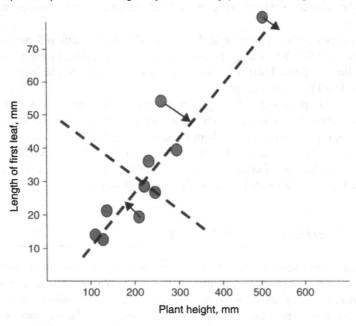

Figure 6.3 An alternative explanation of the operation of PCA, using a two-dimensional dataset (the orchid biometric data plotted in Figure 3.1).

Next, these new axes are rotated so that one axis (the first principal axis) runs along the cloud of data points in such a way as to produce a best fit (Figure 6.3). The term 'best fit' has a precise meaning, similar to the best-fit lines used in multiple linear regression but differing in one important respect. For each data point in turn, draw the residual from the point to the first ordination axis in such a way that the two lines meet at an angle of 90 degrees.

Just as in MLR, the 'best-fit' line is the one which gives the lowest possible sum of squared residuals. It can be shown that this is equivalent to rotating the data points to maximise the information content of the new axis. Additionally, the first principal axis will always pass through the mean of the dataset (as will any other principal axis, also any regression line). This procedure is similar to multiple regression techniques (Chapter 3), but differs in one crucial respect. In regression, residuals are always drawn vertically from the best-fit line, whereas in PCA they are orthogonal to it (i.e. at 90 degrees). This is the same as the reduced major-axis fit (Figure 3.8). Figure 6.4 illustrates the difference between MLR and PCA. The underlying mathematics are very different in the two techniques.

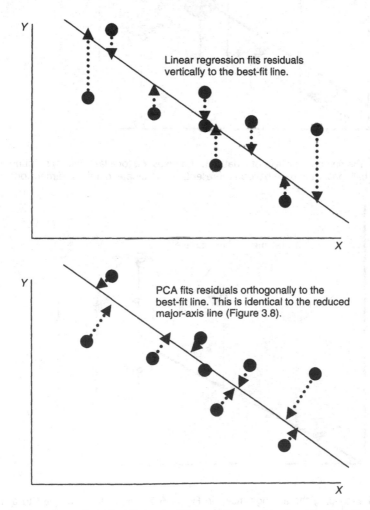

Figure 6.4 A comparison of residual fitting by PCA and linear regression (single or multiple).

An alternative image of a principal axis can be obtained from Figure 6.5, in which each coordinate in a three-variable dataset is shown as a lighter-than-air balloon, and the principal axis is a laser beam shining along between them (specifically the line which encapsulates the greatest variation). This is in contrast to MLR, which would fit a plane in the same situation (Figure 6.6).

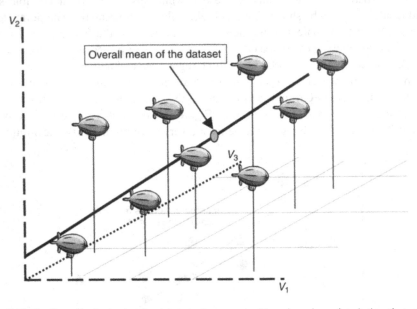

Figure 6.5 The first principal axis of a dataset, shown as rod (or a laser beam) pointing through a set of tethered balloons. Here each balloon represents a coordinate in a three-dimensional data space.

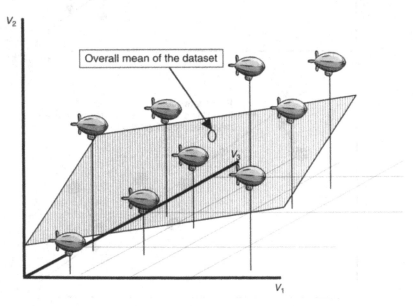

Figure 6.6 Extending the analogy made in Figure 6.5, when MLR is applied to a three-variable dataset it fits a plane.

Having fitted one new axis, the procedure can be repeated. A second 'best-fit' line (the second principal axis) can be fitted exactly as before, with one change. The second ordination axis must satisfy a new condition – it is said to be subject to a constraint. The new axis must meet the first ordination axis at 90 degrees – the lines are said to be orthogonal. For the two-dimensional data in Figure 6.4 this constraint alone is enough to define the second axis. For three-dimensional data one can think of the first axis being the axle of a wheel, with the wheel centred on the mean of the dataset and free to rotate in a complete circle. The second principal axis must lie in the plane of the wheel, and will point along the line of greatest variation (Figure 6.7). The same idea applies in 4D space upwards, but is harder to visualise or draw.

Having fitted a second principal axis, it will usually be possible to fit a third principal axis. Again this axis is constrained; it must meet both the first and second axes at 90 degrees. Generally, for N variables, defining N-dimensional data space, N principal axes can be fitted. Note that the last axis is defined by default – it has only one orientation left available to it. Each successive axis explains less variance than its predecessors, and can generally be assumed to be less important. In practice it is unusual to go beyond the first three principal axes in the interpretation of environmental data.

This (formally correct) interpretation of PCA shows that three types of information can be derived. There is the position of each data point on the new derived axes, the angle of each axis with respect to the pre-existing axes (i.e. the rotations needed to locate each axis), and the amount of variance accounted for by each axis.

With a little thought it should be possible to see that these two explanations of principal components analysis are equivalent. The 'most useful shadow' of the dataset is the scatter-plot generated by the first two principal axes of the data – the most useful ordination diagram for the data. This provides the best available reduction of the data because the first

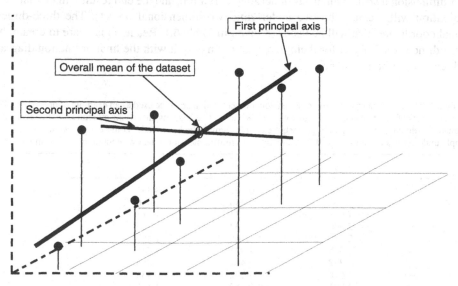

Figure 6.7 Fitting the second principal axis of a dataset. This next axis passes through the mean of the data and is orthogonal (at 90 degrees) to the first axis, but within these constraints it is rotated to find the angle which explains the greatest degree of variation.

principal axis accounts for the greatest possible percentage of the overall variance, and the second axis accounts for the greatest possible proportion of the remaining variance. A shadow is a projection onto a plane, while each principal axis is a projection of a dataset onto a line (hence any two principal axes define a plane).

Both explanations indicate that a PCA will give three useful sets of information about a dataset: projections onto new coordinate axes (i.e. a new set of variables encapsulating the overall information content), the rotations needed to generate each new axis (i.e. the relative importance of each old variable to each new axis), and the actual information content of each new axis.

6.3 Deeper into the mechanics of PCA

6.3.1 About this section

This section is where some readers may start to find the explanations slightly heavy going. It would be possible to skip this entirely, and move on to Section 6.4 if the intention is merely to interpret the output of a PCA. Even so, it would repay the effort of working through the next few pages in order to deepen your understanding of PCA.

This section will work through some of the actual calculations involved in performing a PCA, although it will not attempt to prove why this approach works. The proof involves matrix algebra of a depth that is outside the scope of this book. If needed, the proof is given in most standard textbooks on multivariate analysis.

In order to show exactly what is being done, the process will be worked through by hand using a small dataset which represents a three-dimensional object. This object is in fact a two-dimensional shape which has undergone a rotation, and the end result will be that PCA ordination will recreate the two underlying two-dimensional objects. The three-dimensional coordinates that will be used are given in Table 6.1. Readers may care to create this three-dimensional object for themselves, and compare it with the final ordination diagram (which is given in Figure 6.9).

Table 6.1 The coordinates of the three-dimensional object which are to be used as an example of ordination. This object consists of a two-dimensional figure which has been rotated, and its underlying structure will be revealed by ordination. Although this is a purely geometric example (which the keen reader could construct), it shows how a simple underlying structure can be difficult to detect in multidimensional data without graphical examination.

X	Y	Z
1.716	−0.567	0.991
1.760	−0.480	1.016
1.933	−0.134	1.116
2.366	0.732	1.366
2.582	1.165	1.491
3.015	2.031	1.741
3.232	2.464	1.866
1.616	1.232	0.933
1.991	0.982	1.150
2.741	0.482	1.582
3.116	0.232	1.799

6.3.2 Normalising the data

The first point to consider is that most multivariate datasets consist of extremely differ-ent variables. Plant percentage cover will range from 0 to 100%, with many species being mainly 0%. Animal population values may exceed 10 000 (i.e. insects per square metre), chemical concentrations may take any positive value, and physical properties such as slope or evapotranspiration contain both positive and negative values in one dataset. How can such disparate types of data be made comparable?

The usual approach involves calculating the mean (μ) and standard deviation (s) of each variable separately, then converting each observation X_i into a corresponding Z score Z_i as follows:

$$Z_i = \frac{X_i - \mu}{s} \tag{6.1}$$

The Z score is thus the number of standard deviations each observation is from its own mean. It is dimensionless (i.e. has no units), and by performing this simple transformation each col-umn of data has been converted into a new variable which preserves the shape of the original data but has $\mu = 0$ and $s = 1$. (It is possible to undertake a PCA without this transformation, but the subsequent interpretation is more complicated since the results will be strongly biased towards the variable with the largest numerical values.) This process of conversion to Z scores is known as normalisation, as the well-known normal (Gaussian) distribution is usually han-dled in terms of Z scores. (Note that the 'normalised' data retain their original shape; this transformation does not make data any closer to the classic bell-shaped curve.)

The normalisation of the sample data from Table 6.1 is given in Table 6.2. Visually, the effect of normalisation is to move the mean of the dataset from (μ_1, μ_2, ...) to (0, 0, ...) and to rescale the axes so that all are measured in equivalent units but to leave the shape unchanged.

Table 6.2 The normalisation of the data from Table 6.1. This operation is also known as centring the data. Notice that each observation is converted to a value indicating how many standard deviations it is from its mean, and that after normalisation the mean becomes 0.0 and the standard deviation 1.0. Thus the first X value is trans-formed as:
New X score = (1.716 − 2.370)/0.600 = −1.09, etc.

	Before normalisation			After normalisation		
	X	Y	Z	X	Y	Z
	1.716	−0.567	0.991	−1.09	−1.35	−1.09
	1.760	−0.480	1.016	−1.02	−1.26	−1.02
	1.933	−0.134	1.116	−0.73	−0.90	−0.73
	2.366	0.732	1.366	−0.01	−0.01	−0.01
	2.582	1.165	1.491	0.35	0.44	0.35
	3.015	2.031	1.741	1.08	1.33	1.08
	3.232	2.464	1.866	1.44	1.78	1.44
	1.616	1.232	0.933	−1.26	0.51	−1.26
	1.991	0.982	1.150	−0.63	0.25	−0.63
	2.741	0.482	1.582	0.62	−0.27	0.62
	3.116	0.232	1.799	1.24	−0.52	1.24
mean:	2.370	0.740	1.368	0.00	0.00	0.00
sd:	0.600	0.970	0.346	1.00	1.00	1.00

6.3.3 The extraction of principal components

Now comes the difficult bit! Fortunately this would normally be undertaken by computer, although for 2D data the result may be derived by hand. The cloud of N-dimensional data points needs to be rotated to generate a set of N principal axes. To do this involves finding a set of numbers which indicate the amount of rotation applied to each variable. These numbers are called loadings (or alternatively weightings). The ordination is achieved by finding a set of loadings which rotates the data to give the best possible fit to the data.

How do we find the best possible values for the loadings given to each variable? And how are loadings found for the second, third and higher principal axes given the additional constraints that they must be at 90 degrees to all previous axes? Random searching is out of the question, as the number of possible combinations of loadings is infinite.

The solution to this question is given in any advanced books on the matrix algebra underlying multivariate statistics. *The answer is given by finding the eigenvectors and eigenvalues of the Pearson's correlation matrix.* These terms will need explaining carefully.

6.3.4 The correlation matrix

The Pearson's correlation matrix is simply the matrix of all possible Pearson's correlation coefficients between the variables under examination. This is routinely produced by most statistical packages, and is straightforward (if tedious) to calculate by hand. An example from the three-variable dataset of Table 6.1 is given in Table 6.3.

Note that the correlation matrix has a diagonal consisting entirely of 1s. This diagonal is known as the leading diagonal, and the fact that it is all 1s simply says that the correlation coefficient between each variable and itself is 1.0. Correlation matrices always have a leading diagonal composed wholly of 1s. The rather artificial nature of the data is highlighted by the high X/Z correlation, and the fact that the X/Y and Y/Z correlations are almost identical.

When presenting correlation matrices in publications it is good practice to indicate the significance of each correlation, or simply exclude non-significant results. When generating correlation matrices for PCA all values are used, even when the correlation coefficient is clearly non-significant.

6.3.5 Eigenvalues and eigenvectors

For a full explanation of these terms the reader is referred to any good textbook on matrix algebra. The intention here is merely to give enough of the flavour of their meaning to allow an interpretation of a PCA output.

Table 6.3 The correlation matrix (using Pearson's product moment correlation coefficient) for the sample data given in Table 6.1. No attempt is made to examine the significance levels in this case, since PCA uses the full matrix irrespective of the significance of any individual correlation. Notice that the matrix is symmetrical around the leading diagonal, which must always consist of 1.0s. Notice also that the artificial nature of the data is shown by the extremely high correlation between X and Z, and the fact that the X/Y and Y/Z correlations are very similar.

	X	Y	Z
X	1.000	0.593	0.999
Y	0.593	1.000	0.594
Z	0.999	0.594	1.000

Table 6.4 The basics of matrix multiplication.

A matrix (plural matrices) is a standard way of arranging numerical information concerning linear equations. They are rectangular, and may contain any non-zero number of rows and columns. A matrix with C columns and R rows is referred to as a $(C \times R)$ matrix – notice that this information refers solely to its size, and gives no indication about its contents. By convention matrices are shown enclosed in square brackets. Numbers within matrices are known as their elements, and are described by reference to their position. Thus in a matrix M the top left number may be referred to as $M[1,1]$, etc.

It is possible to multiply two matrices together, providing that the number of columns in the first matrix equals the number of rows in the second. The formal definition of matrix multiplication is as follows: If M_1 is an $(A \times B)$ matrix, it may be multiplied by a $(C \times A)$ matrix M_2, generating a $(C \times B)$ matrix M_3 according to the rule:

$$M_3[x,y] = M_1[1,y] \times M_2[x,1] + M_1[2,y] \times M_2[x,2] + M_1[3,y] \times M_2[x,3] + \cdots$$

This complicated-looking formula may be explained by a simple example involving the costs of different shopping baskets containing various combinations of the same fruits. In this example an apple costs 10 p, an orange 15 p and a banana 25 p. It is trivial to calculate that a basket with one apple, one orange and one banana costs $(1 \times 10 + 1 \times 15 + 1 \times 25) = 50$ p, while one with three apples, two oranges and one banana costs $(3 \times 10 + 2 \times 15 + 1 \times 25) = 85$ p.

In matrix notation this would be written by setting up two matrices, one with the quantities of the fruits and one with their costs.

Quantity matrix = $\begin{bmatrix} 1 & 1 & 1 \\ 3 & 2 & 1 \end{bmatrix}$

Price matrix = $\begin{bmatrix} 10 \\ 15 \\ 25 \end{bmatrix}$

Cost matrix = quantity × price

$$= \begin{bmatrix} 1 & 1 & 1 \\ 3 & 2 & 1 \end{bmatrix} \times \begin{bmatrix} 10 \\ 15 \\ 25 \end{bmatrix} = \begin{bmatrix} 1 \times 10 + 1 \times 15 + 1 \times 25 \\ 3 \times 10 + 2 \times 15 + 1 \times 25 \end{bmatrix} = \begin{bmatrix} 50 \\ 85 \end{bmatrix}$$

showing that the two different permutations of purchases come to a total cost of 50 p and 85 p respectively. Thus matrix multiplication is just a new way of presenting simple and familiar sets of linear calculations.

Eigenvalues and eigenvectors are intimately connected with the operation of matrix multiplication, which is explained in Table 6.4. When a square $(N \times N)$ matrix is multiplied with a $(1 \times N)$ matrix, the result is a new $(1 \times N)$ matrix. This operation can be repeated on the new $(1 \times N)$ matrix, generating another $(1 \times N)$ matrix. Table 6.5 gives an example of this process in action with a (3×3) matrix. After a number of repeats (iterations), the pattern of numbers generated settles down to a constant shape, although their actual values change each time by a constant amount. This may be seen more easily in Figure 6.8, which shows the same information as Table 6.5 but in graphical format. This effect, whereby repeatedly applying the same matrix multiplication to successive generations of numbers produces a stable pattern, is true for any square $(N \times N)$ matrix, and the pattern that emerges is independent of the initial values used for the $(1 \times N)$ matrix. It is a standard result that repeatedly applying the same matrix multiplication causes any initial set of numbers to behave in the same general way. The numbers settle down towards a system where they will all grow (or shrink) at a constant rate, and their relative values settle down to a

Table 6.5 An example of repeatedly applying the same matrix multiplication to successive generations of numbers, to illustrate the way in which this process results in a stable pattern. In this example the matrix used is the matrix of correlation coefficients given in Table 6.3, and the process starts by multiplying a (1×3) matrix of 1s. The final pattern that results would be the same for almost any (1×3) matrix. Such repeated calculations are known as iterative, and each successive generation represents one iteration.

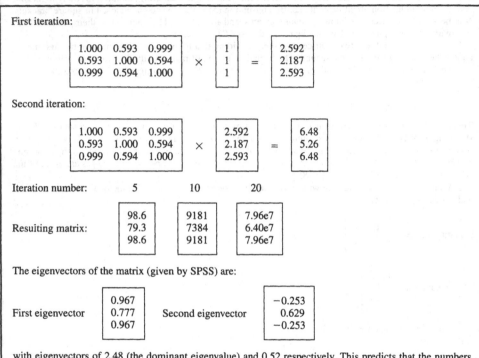

First iteration:

$$
\begin{bmatrix} 1.000 & 0.593 & 0.999 \\ 0.593 & 1.000 & 0.594 \\ 0.999 & 0.594 & 1.000 \end{bmatrix} \times \begin{bmatrix} 1 \\ 1 \\ 1 \end{bmatrix} = \begin{bmatrix} 2.592 \\ 2.187 \\ 2.593 \end{bmatrix}
$$

Second iteration:

$$
\begin{bmatrix} 1.000 & 0.593 & 0.999 \\ 0.593 & 1.000 & 0.594 \\ 0.999 & 0.594 & 1.000 \end{bmatrix} \times \begin{bmatrix} 2.592 \\ 2.187 \\ 2.593 \end{bmatrix} = \begin{bmatrix} 6.48 \\ 5.26 \\ 6.48 \end{bmatrix}
$$

Iteration number:	5	10	20
Resulting matrix:	98.6	9181	7.96e7
	79.3	7384	6.40e7
	98.6	9181	7.96e7

The eigenvectors of the matrix (given by SPSS) are:

First eigenvector $\begin{bmatrix} 0.967 \\ 0.777 \\ 0.967 \end{bmatrix}$ Second eigenvector $\begin{bmatrix} -0.253 \\ 0.629 \\ -0.253 \end{bmatrix}$

with eigenvectors of 2.48 (the dominant eigenvalue) and 0.52 respectively. This predicts that the numbers given above should stabilise to the ratios of 967:777:967, and increase by a factor of approximately 2.48 with each generation.

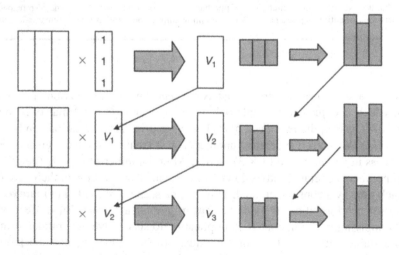

Figure 6.8 The growth of an eigenvector after repeated matrix multiplication. After a while, each successive multiplication preserves the shape (the eigenvector) while increasing values by a constant amount (the eigenvalue).

constant pattern. The rate of growth (or shrinkage) per multiplication is known as the dominant eigenvalue, and the pattern they form is the dominant (or principal) eigenvector.

Table 6.5 gives the dominant eigenvector of the square matrix used (supplied by the package SPSS for Windows). It can be seen that this calculated eigenvector is indeed the pattern to which the system stabilises. The dominant eigenvalue for the matrix is 2.48, meaning that once equilibrium is reached each generation of numbers increases by a factor of 2.48.

Formally, an eigenvector V of a square matrix M can be defined as follows:

$$M \times V = \lambda \times V \tag{6.2}$$

where λ is the eigenvalue. Note that M is a square ($N \times N$) matrix, V is a ($1 \times N$) matrix, but λ is just a single number. In matrix algebra, single numbers tend to be called scalars to emphasise the fact that they are not matrices. Hence equation 6.2 simply states that post-multiplying a matrix by one of its eigenvectors is equivalent to a scalar multiplication of the eigenvector.

For simple matrices such as this, eigenvalues and their associated eigenvectors can be derived by hand. Generally one uses computers, and I am going to make no attempt to explain the manual derivation. (In fact modern software often performs this and other ordinations by iterative techniques which calculate eigenvalues, eigenvectors and axis scores in parallel, since these are more computationally efficient.)

It is a standard result that a square ($N \times N$) matrix has N eigenvectors, each with an associated eigenvalue. The importance of each eigenvector is given by the size of its associated eigenvalue, with progressively smaller eigenvalues denoting progressively less importance.

6.3.6 The relationship between eigenvectors and PCA

The best way to understand how the eigenvectors relate to PCA ordination is to appreciate that PCA is about rotating data. PCA takes a set of R observations on N variables as a set of R points in an N-dimensional space. A new set of N principal axes is then derived, each one defined by rotating the dataset by a certain angle with respect to the old axes (Figure 6.7).

The shape of the pattern of points in the new space is unchanged, but has been rotated to allow the operator the most useful view of the dataset. The value of the operation is that the first axis of the new space (the first principal axis of the data) encapsulates the maximum possible information content, the second axis contains the second greatest information content and so on.

Precisely *why* eigenvectors and eigenvalues are related to ordination techniques is outside the scope of this book. It is sufficient to think of them as a powerful way of summarising the underlying structure of a matrix, and to accept that the eigenvectors of certain types of matrix which gives measures of relatedness or distance between a set of variables can give a useful summary of the structure of the underlying data. In the case of PCA the ordination relies on the matrix of correlation coefficients; other ordination techniques use matrices containing other measures of the similarity between variables. Thus principal coordinates analysis (Chapter 10) uses eigenvectors of a matrix summarising distances between data points, and correspondence analysis (Chapter 7) involves eigenvectors of a matrix of χ^2 values.

6.3.7 Why different packages can give different eigenvectors

There remain a few important properties of eigenvectors that need to be understood in order to interpret the output of a PCA (or other ordination). Firstly, eigenvectors can best be thought of as a shape – a relative pattern of numbers which is preserved under matrix multiplication. This is illustrated in Figure 6.9.

Note that this result is independent of the actual values given to each element. The axis scale in Figure 6.9 could be multiplied by 10, or 100, or $-\frac{1}{2}$, or any other number and the picture shown would remain valid. This means that for any calculated eigenvector, there are infinitely many equally valid sets of numbers that could also be described as the same eigenvector. (They would all be simple multiples of each other, and within each the numbers would all show the same *relative* sizes.) There are conventions for standardising eigenvectors (by imposing another constraint, such as ensuring that the first element is 1, or that the sum of their elements squared equals a set value such as 1.0) but these involve essentially arbitrary decisions.

This insight into eigenvectors has a very important consequence for any eigenvector-based ordination technique. This is that the actual numbers produced by one computer package may differ greatly from the output from another package – without either package being wrong. Comparing the outputs from three packages offering PCA ordination (two commercial and one written by the author), the values did indeed differ (by a factor of 2 in one case, and by −1 in another). All are equally correct – this can be checked by plotting a graph of the ordination results, when the patterns should be found to be identical in all cases. (See Section 6.6 for techniques to explore the results of a PCA.) This contrasts

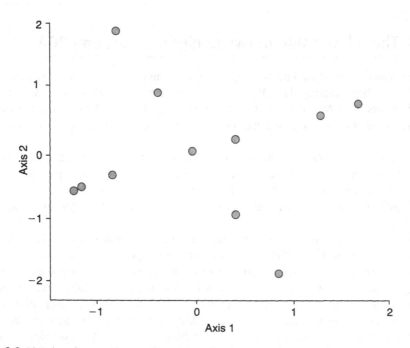

Figure 6.9 The data from Table 6.1 after PCA ordination, revealing that the points defined a cross shape.

sharply with standard univariate statistical techniques, where one would immediately have grave doubts if two packages gave differing values for a standard deviation or an F score.

It was explained in Section 4.1 that ordinations should be thought of as being drawn on a sheet of glass. They can be rotated or flipped over while remaining valid. In the case of PCA this arises directly out of the fact that an eigenvector may be multiplied by -1 and remains a valid eigenvector.

6.3.8 The interpretation of eigenvalues

A second important result comes from the eigenvalues. These give a precise indication of the relative importance of each ordination axis, with the largest eigenvalue being associated with the first principal axis, the second largest eigenvalue being associated with the second principal axis, etc. The eigenvalues of a dataset should not differ appreciably between packages (unlike the eigenvectors). A matrix with 20 species would generate 20 eigenvectors, but in practice it is unlikely that any except the first three or four would be of any importance for interpreting the data. The relation between eigenvalues and variance is simple to state for PCA:

$$V_m = \frac{100 \times \lambda_m}{N}$$

(6.3)

where V_m is the percentage variance explained by the mth ordination axis, λ_m is the mth eigenvalue, and N is the number of variables. Another way to state this result is that the eigenvalues of an $N \times N$ correlation matrix will always sum to N. (This simple relationship does not hold true for other ordination techniques involving eigenvectors/values. Most packages will supply a percentage variance along with any eigenvalue, along with a cumulative total.)

Curiously, there is no formal test of significance available to decide whether any given ordination axis is meaningful or just random noise, nor is there any test to decide whether or not individual variables contribute significantly to an ordination axis. PCA, like all ordinations, is a summary description of data, not an inferential test. Ways to circumvent this limitation are described in Section 6.5.

6.3.9 Derivation of axis scores

It is now possible to explain the derivation of the principal axis scores, from which the ordination diagrams are derived. These come from the normalised data (Section 6.3.2) and the eigenvectors (Section 6.3.4). The Nth axis is derived by multiplying the matrix of normalised data by the Nth eigenvector. This is shown in Table 6.6, in which the first two axis scores are calculated for the data of Table 6.1. These data are graphed in Figure 6.9, which shows that the 3D data of Table 6.1 in fact portrayed the letter X, rotated in three dimensions.

6.4 Preliminary interpretation of a PCA

In this section we will explore a simple example of a PCA ordination, and use it to rehearse the standard ways to portray and interpret the results. Actually running the PCA is quick and simple on suitable packages; the hard part comes in interpreting the output!

Table 6.6 The derivation of the principal axis scores for the data given in Table 6.1.

The first principal axis scores are derived by multiplying the normalised data (given in Table 6.2) by the first (dominant) eigenvector:

$$
\begin{bmatrix}
-1.09 & -1.35 & -1.09 \\
-1.02 & -1.26 & -1.02 \\
-0.73 & -0.90 & -0.73 \\
-0.01 & -0.01 & -0.01 \\
0.35 & 0.44 & 0.36 \\
1.08 & 1.33 & 1.08 \\
1.44 & 1.78 & 1.44 \\
-1.26 & 0.51 & -1.26 \\
-0.63 & 0.25 & -0.63 \\
0.62 & -0.27 & 0.62 \\
1.24 & -0.52 & 1.25
\end{bmatrix}
\times
\begin{bmatrix}
0.967 \\
0.777 \\
0.967
\end{bmatrix}
=
\begin{bmatrix}
-3.16 \\
-2.95 \\
-2.11 \\
-0.02 \\
1.02 \\
3.12 \\
4.17 \\
-2.04 \\
-1.02 \\
0.99 \\
1.99
\end{bmatrix}
$$

Similarly, the second axis scores are given by the second eigenvector:

$$
\begin{bmatrix}
-1.09 & -1.35 & -1.09 \\
-1.02 & -1.26 & -1.02 \\
-0.73 & -0.90 & -0.73 \\
-0.01 & -0.01 & -0.01 \\
0.35 & 0.44 & 0.36 \\
1.08 & 1.33 & 1.08 \\
1.44 & 1.78 & 1.44 \\
-1.26 & 0.51 & -1.26 \\
-0.63 & 0.25 & -0.63 \\
0.62 & -0.27 & 0.62 \\
1.24 & -0.52 & 1.25
\end{bmatrix}
\times
\begin{bmatrix}
-0.253 \\
0.629 \\
-0.523
\end{bmatrix}
=
\begin{bmatrix}
-0.30 \\
-0.28 \\
-0.20 \\
0.00 \\
0.10 \\
0.29 \\
0.39 \\
0.96 \\
0.48 \\
-0.48 \\
-0.95
\end{bmatrix}
$$

6.4.1 Example

We will start by running a PCA on the Wimbledon Common dataset (Table 1.4). It is immediately obvious that there are two distinct communities; heathland (with heather *Calluna vulgaris*, wavy hair grass *Deschampsia flexuosa*, and red fescue *Festuca rubra*) and clay spoil (with false oat grass *Arrhenatherum elatius*, hogweed *Heracleum sphondylium*, yarrow *Achillea millefolia*, and the legumes *Trifolium repens* and *Vicia sativa*). In addition to these floral variables, there are two 'classification' variables; the habitat type and the replicate number. Notice that these two classification variables are not included in the ordination, but are reserved for subsequent statistical analyses.

This analysis will be performed on the data for the plant species only. It would be mechanically possible to include the classification variables in the PCA, but this must be avoided as completely incorrect for several reasons. Firstly this would involve mixing two fundamentally different datasets containing different types of data, and as such would be unlikely to give a useful description of either dataset. Secondly including the metadata in a PCA would preclude the possibility of using these metadata for any formal testing on the outputs from the PCA.

From the explanation given above, a PCA performed on eight variables will produce eight new principal axes of progressively lesser importance. In this case we will start off by listing the output for the first six axes, although in practice few researchers look beyond the fourth axis of a PCA, and in this case we will find that only the first two need be considered.

The exact layout of a PCA output, and what information it lists, varies between packages. The minimum output must include listings of the eigenvalues for each axis (usually with their corresponding percentages of variance using equation 6.3), plus the eigenvector loads (also called factor loadings) for each variable on each axis. The package must also be able to calculate the axis scores for each observation on each axis, although it should be noted that not all packages automatically create the axis scores as a default, and the user may have to set a 'save as variables' option within the PCA.

These three sets of information for the Wimbledon Common data are listed in full in Table 6.7. This shows that the first axis accounts for 53.1% of the total variance, with 26.2% on the second axis. These eigenvalues should be approximately the same whichever package is used (although minute deviations may arise from cumulative errors after long floating-point calculations).

The raw PCA data listed in Table 6.7 will now be used to explain the procedures by which one interprets a PCA. This is largely an exercise in hypothesis generation, although it should be followed by hypothesis testing where possible.

Hypothesis generation is perceived by many to be the main use of multivariate statistics, but always carries an element of subjectivity. It involves 'eyeballing' the three different aspects of the output listed in Table 6.7. The recommended sequence of operations is: (1) examination of percentage variance, (2) inspection of eigenvector loadings, and (3) plotting of ordination diagrams. If a testable hypothesis results, it should be examined using inferential statistical tests on the new ordination axes (described under Section 6.4.6).

6.4.2 Examination of percentage variance

The first information to check is the percentage variance explained by the first few principal axes. (In case this is not given, an eigenvalue will be, and the relation between the two is given in equation 6.3.) The larger the variance, the greater the amount of information that has been condensed into the ordination axis. In the case of totally shapeless data each axis will account for the same amount of the variance, equal to $100/N\%$ where N is the number of variables in the dataset. In practice, however, even datasets composed of random numbers are not perfectly shapeless (Richman, 1988), and unfortunately there are no formal tests available to establish significance.

The simplest test, supplied as a default option on many packages, is to reject any axis whose eigenvalue is less than 1 (i.e. its percentage variance is less than $100/N$, where N is the number of variables). This is known as the Kaiser–Guttmann criterion (Cliff, 1988). Jackson (1993) reviewed a variety of approaches to deciding the importance of principal axes from their eigenvalues, and concluded that the Kaiser–Guttman criterion tended to overestimate the number of non-trivial axes. He suggested that the broken stick model was the most suitable approach to deciding which principal axes should be examined. This model suggests that for random data with N variables, the pth axis should have an eigenvalue λ_p as follows:

$$\lambda_p = \sum_{i=n}^{i=p} (1/i)$$

Some broken stick values are given in Table 6.8 (recoded as percentage variance), but other values are straightforward to calculate if needed. If the observed percentage variance is

Table 6.7 The outputs from a PCA on the Wimbledon common data of Table 1.4.

The eigenvalues and percentage variance
The eigenvector loadings (excluding axes 6, 7 and 8, which contain negligible
information)

Axis	Eigenvalue	%variance	cumulative variance
1	4.248	53.096	53.096
2	2.096	26.199	79.295
3	0.8	10.005	89.3
4	0.332	4.147	93.446
5	0.3	3.752	97.198

The eigenvector loadings (species loadings or factor loadings). Species
abbreviations follow Table 1.4

Species	axis 1	axis 2	axis 3	axis 4	axis 5
Am	0.3479	-0.3771	-0.0771	0.304	-0.6807
Ae	0.4686	-0.1127	-0.113	0.1224	0.2112
Fr	-0.4163	-0.0558	-0.4252	-0.4318	-0.1822
Cv	-0.3616	-0.0416	0.6723	0.4069	0.0696
Df	-0.3617	-0.0542	-0.5743	0.6914	0.2196
Hs	0.2255	0.581	-0.0731	0.0118	0.249
Tr	0.3081	-0.4848	-0.0543	-0.0545	0.5482
Vs	0.2806	0.5146	-0.1021	0.2438	-0.2082

The axis scores

Site	rep	axis 1	axis 2	axis 3	axis 4	axis 5
Heath	1	-1.99192	-0.13024	-1.50671	-0.20891	-0.01157
Heath	2	-1.30253	-0.01010	2.01062	0.01634	0.00430
Heath	3	-2.81910	-0.22881	-0.19468	1.07278	0.27846
Heath	4	-1.94687	-0.10665	0.07706	-1.08918	-0.3764
Mound	1	2.27756	-1.66780	-0.09759	0.36361	-1.04025
Mound	2	2.25224	-2.15231	-0.05306	-0.25520	0.98712
Mound	3	1.97343	1.99709	-0.17395	0.23776	-0.15819
Mound	4	1.55719	2.29881	-0.06169	-0.13721	0.31655

greater than the expected value, the axis may be expected to contain some interpretable information.

Examining Table 6.8 for a dataset with eight variables, we find that the broken stick model gives predicted values of 33.9%, 21.4% and 15.2% variance for axes 1–3, while our actual data shows values of 53.1%, 21.6% and 10.0%. On this basis the first two axes seems worthy of investigation, while the third (and higher) axes do not.

6.4.3 Eigenvector loadings

We confine our attention to the first two axes (following comparison of the outputs with the broken stick model). The eigenvector loadings for the first axis are given in part 2 of Table 6.7 and graphed in Figure 6.10, showing negative loadings to the heathland species and positive loadings for spoil mound species. This suggests that the first axis is related to the difference

Table 6.8 Expected values for the percentage variance accounted for by the first five principal axes of a PCA under the broken stick model. Axes found to account for less variance than the values tabulated here may be disregarded as random noise.

Number variables	axis 1	axis 2	axis 3	axis 4	axis 5
3	61.1	27.7	11.1	–	–
4	52.0	27.0	14.5	6.2	–
5	45.6	25.6	15.6	8.9	3.9
6	40.8	24.1	15.8	10.2	6.1
7	37.0	22.7	15.6	10.8	7.2
8	33.9	21.4	15.2	11.0	7.9
9	31.4	20.3	14.7	11.0	8.2
10	29.2	19.2	14.2	10.9	8.4
11	27.4	18.3	13.8	10.7	8.5
12	25.8	17.5	13.3	10.5	8.4
13	24.4	16.7	12.9	10.3	8.4
14	23.2	16.0	12.5	10.1	8.3
15	22.1	15.4	12.1	9.8	8.2
16	21.1	14.8	11.7	9.6	8.1
17	20.2	14.3	11.4	9.4	7.9
18	19.4	13.8	11.0	9.2	7.8
19	18.6	13.4	10.7	9.0	7.7
20	17.9	12.9	10.4	8.8	7.5
25	15.2	11.2	9.2	7.9	6.9
30	13.3	9.9	8.3	7.2	6.3
35	11.8	8.9	7.5	6.6	5.8
40	10.6	8.1	6.9	6.1	5.4
50	8.9	6.9	5.9	5.3	4.8
60	7.7	6.1	5.2	4.7	4.3
70	6.9	5.4	4.7	4.2	3.9
80	6.2	4.9	4.3	3.9	3.6
90	5.6	4.5	3.9	3.6	3.3
100	5.1	4.1	3.6	3.3	3.1
150	3.7	3.0	2.7	2.5	2.3
200	2.9	2.4	2.1	2.0	1.8

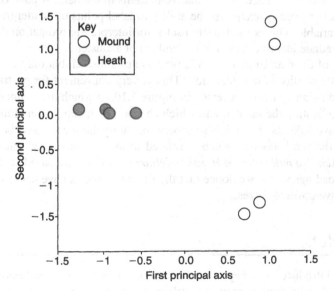

Figure 6.10 A graph showing the first two axes of the PCA ordination of the Wimbledon Common data (Table 1.4).

between the habitats, as would be expected from the nature of the data. It must be emphasised that different packages may give very different-looking results for these eigenvector loadings, although the underlying pattern should be the same. (See Section 6.4 for the explanation of this.) The pattern of eigenvector loadings on the second axis is dominated by negative loadings for *Trifolium repens* and *Achillea millefolia*.

The next step is to examine an ordination diagram.

6.4.4 Plotting of ordination diagrams

The PCA has created a new set of variables, corresponding to the position of each observation on each of the principal axes. If your dataset contained N variables you could potentially generate N new principal axes, although in practice it is normal that only the first 2–4 are of any use. Some packages need to be prompted to produce these values, and there is no standardisation about the names given to the new variables. Typical names are PRIN_n, AXIS_n or FAC_n (for the nth axis scores in each case).

These new variables can legitimately be handled as conventional data. In particular, it is extremely useful to plot them out as scattergraphs. This is equivalent to casting the shadow of the N-dimensional dataset, as described in Section 6.2. The resulting scattergraph is an ordination diagram (as described in Chapter 4). In principle these ordination diagrams could be plotted by hand – this is tedious, and will almost always be performed by the same package that undertook the ordination.

The first graph to plot is certainly the scattergraph of axis 1 against axis 2, since this must hold the greatest percentage of the overall variance. It is not unusual for the first two axes together to account for 90% of the total variance in a multivariate dataset, so that the majority of the information content of the data can be seen in one graph. Other combinations of axes should also be inspected, such as axis 1 against axis 3. In each case the objective is the same: to seek visual evidence of informative patterns in the data. A powerful and helpful feature of most graphics packages is the ability to label points on scattergraphs according to a marker variable. This is immensely useful for interpreting ordination diagrams, since it allows immediate identification of the members of a cluster.

In the case of the data from Table 6.7, we have established that only the first two axes are likely to be worthy of consideration. This is very convenient for portrayal by a two-dimensional diagram! This is presented in Figure 6.10, in which the ordination is overlain by symbols indicating the habitat, and which shows a clear separation along the first axis between the two habitats. There is also separation along the second axis between the data collected on the spoil mound (which is related to the presence or absence of *Trifolium repens* with *Achillea millefolia*, or *H. sphondylium* with *V. sativa*, based on the second axis eigenvector loadings where we notice that the first two species have negative scores while the second have positive scores).

6.4.5 Biplots

Biplots were introduced in Chapter 4 as an extension of the basic ordination diagram, in which a second ordination diagram is overlain on top of the first in order to show relationships between the two sets of data. It was explained that the actual implementation of the second ordination varied according to the technique used. These principles will now be exemplified using PCA ordination.

The biplot arising from a PCA consists of two sets of information: a scattergraph of ordination scores for the rows of the matrix (i.e. the standard PCA ordination diagram), and a graphical representation of the corresponding set of loadings given to each variable (the eigenvector loadings). As is usual with biplots, the two types of data are plotted differently. Ordination scores are shown as conventional coordinates, while eigenvector loadings are shown as an arrow leading from the origin to the designated coordinate. Biplots are not a substitute for statistical analysis; biplots suggest tendencies but never prove anything. (In fact it would be incorrect to attempt to use a standard inferential statistical test to examine the relationship between the two sets of information given in a biplot, since by definition they are not independent. Section 6.4.6 enlarges on this point.)

The eigenvector loadings are added as arrows in Figure 6.11 to generate a biplot summarising the ordination. The arrows representing the heathland species are effectively superimposed, pointing left towards the heathland sites. The species found on the clay spoil site all point right, but while *Arrhenetherum elatius* points straight right, hogweed *Heracleum sphondylium* and *Vicia sativa* point up (positive loadings on the second axis) while *Achillea millefolia* and *Trifolium repens* point downwards. This can be interpreted as follows: The biggest separation in the data is between the heathland and the clay spoil sites, indicated by their separation on the first principal axis. The second greatest source of variation is within the spoil mound habitat, with *Arrhenetherum elatius* occurring in all plots but *Heracleum sphondylium* and *Vicia sativa* only co-occurring in two plots. (A reexamination of Table 1.4 confirms that this is exactly what the source data show.)

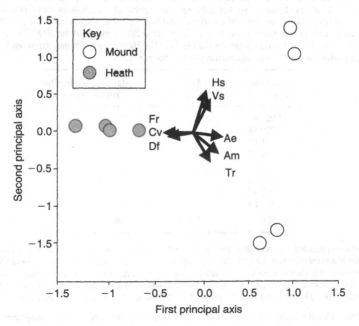

Figure 6.11 The PCA ordination of the Wimbledon Common data converted into a biplot by addition of arrows showing species loadings. Abbreviations follow Table 1.4.

6.4.6 Applying inferential statistics to ordination data

The approaches described above to interpret a principal components ordination can be criticised as being vague and subjective. Fortunately they can be backed up by rigorous inferential statistics, although this step is often omitted from published ordinations. The point was made in Section 6.4.5 that the principal axis scores can be handled in the same way as primary (i.e. measured) variables. In addition to being plotted, they can be subjected to regression analysis or analysis of variance, subject to two conditions.

The first condition for applying inferential statistics to ordination scores concerns the usual criterion of normality. Parametric tests (such as Pearson's correlation coefficient or ANOVA) require that the data analysed conform to a normal distribution. Ordination scores should be checked for normality by standard techniques (Section 1.5.4), although in fact they are often a better approximation to the normal distribution than the raw data from which they were derived (especially when the input to the PCA consists of a large number of non-normal variables such as questionnaire returns). Table 6.9 includes the result of running a standard

Table 6.9 Statistical analyses of ordination scores from the PCA in Table 6.7.

ANOVA on first axis score by habitat:
$F_{1,6} = 130.3, p < 0.01$ according to standard models.
However, applying a Monte Carlo test shows that this significance is over-stated:

An example of a Monte Carlo simulation to provide significance levels for a standard parametric test on ordinated data will be described. Such a procedure is laborious but provides an accurate significance level in cases where the standard assumption of normality may be invalid. The data in this example are the first axis scores of the PCA in Table 6.7, broken down by habitat.

Columns one and two give the actual axis scores, and the habitat to which the score applies. This is used as the classification variable in a standard one-way ANOVA, producing an F value of 130.3. The column defining habitat is now randomised, so that although the number of each habitat remains unchanged their order bears no relationship to the true data, and the ANOVA repeated. This generates a new F value. This randomization procedure is repeated many times – in this case 2000 times, and the 2000 F values sorted into ascending order. The actual F value is greater than 95% of the 2000 values obtained from random sequences of the data, showing the result to be significant (at least for $p < 0.05$).

First axis score	True habitat	Randomised habitat (First iteration)
−1.99192	H	H
−1.30253	H	S
−2.81910	H	S
−1.94687	H	S
2.27756	S	H
2.25224	S	S
1.97343	S	H
1.55719	S	H
F value:	130.1	1.6

The randomisation process was repeated 2000 times, and the resulting F values sorted in ascending order. The 95% value for F was therefore the (0.95×2000)th largest value, which was found to be 3.1, while the 99% value was found to be 130.1. Consequently the value of F found for the real data may be accepted as significant, but only at $p < 0.05$. (Note that this Monte Carlo test has shown that the apparently huge F value of 130 is expected to occur one time in 100 with these data, far more often than would be expected in normally distributed data.)

Beware: Pearson's correlation coefficient between the first axis score and the percentage cover of *Calluna vulgaris* is $r_6 = -0.75, p < 0.01$, *but* this analysis is invalid as the variables are not independent.

one-way analysis of variance on the first axis scores for the Wimbledon Common dataset (Table 6.7) testing the null hypothesis that there is no difference between the first axis scores for each habitat, showing an apparently highly significant F value of 130.3. (Although PCA scores may differ between packages, the F score resulting from applying ANOVA to them should be the same whichever ordination package is used. However, deciding on their significance is more difficult.)

In the event that ordination scores appear to be substantially non-normal, inferential tests may still be applied in two different ways. The simplest solution is to use non-parametric tests such as Spearman's correlation coefficient or Kruskal–Wallis ANOVA. These rely solely on the relative ordering of data points without making any assumptions about their distribution, and are widely used although slightly less powerful than their parametric equivalents.

An alternative (and far preferable) approach is to run a Monte Carlo test (Manly, 1991), which essentially involves comparing the results of a test on your data with the results of the same test applied to hundreds of randomised versions of the same data. The technique is as follows: Perform your chosen inferential test on your ordination scores – this could involve correlating them with an environmental variable, classifying them then performing ANOVA, etc. Record the test statistic that is produced. If the test were an ANOVA producing an F statistic, call this F^*. Now randomise the ordering of the data (in this case randomise the ordering of observations with respect to habitats) and repeat the analysis. Note the resulting test statistic. Repeat the randomisation/reanalysis procedure at least 200 times, and sort the resulting set of test statistics into ascending order. This gives an unbiased indication of the expected distribution of the test statistic arising from your data. Now compare your actual value (F^*) with this distribution. If it exceeds the centile corresponding to your chosen level of significance, you may reject the null hypothesis and accept that there is a pattern in your ordination scores. Thus if your criterion for significance is $p < 0.05$ and 1000 simulations are run, the critical value of F would be the 950th largest value obtained. An example is given in Table 6.9, where we apply a Monte Carlo test, reshuffling, then reanalysing the data 2000 times, and make the surprising observation that this F value is in fact only significant at $p < 0.05$, occurring about 1 time in 100 with permutations of these data. One issue that currently limits the usage of Monte Carlo tests is the relative shortage of software which offers this model of inferential testing. The permutations used in Table 6.9 were performed by a small customised program written by the author.

The second criterion for applying inferential statistics to ordination scores is also standard, but provides a potential pitfall in multivariate situations. This concerns the requirement that there should be no a priori connection between dependent and independent variables. The catch is that an ordination score is composed of a set of variables, each given a weighting then added together. (Section 6.3 gives more details on this.) The resulting ordination scores *cannot* therefore be assumed to be independent of any of its constituent variables. This cannot be bypassed by any transformation nor by Monte Carlo simulations since it requires care in the setting up of the actual ordination. Consequently when planning to analyse PCA output (or any other type of ordination score) with respect to some classification variable, it is crucial that the classification variable be excluded from the ordination. Table 6.9 illustrates this pitfall with a deliberately invalid correlation as an example of bad practice. The first axis scores correlate significantly with percentage cover of *Calluna vulgaris*, but this result is in fact worthless since *Calluna* was used to derive the ordination scores in the first place. It would also be possible with a minor recoding of the data to obtain a *t* test or ANOVA comparing axis scores for sites where a species occurred with scores for sites without this plant. Such a comparison would be equally invalid, again because the axis score would not be independent of the source data. This is why environmental variables should be excluded from a species ordination, so that they could validly be used in subsequent tests.

6.5 Other worked examples of PCA

6.5.1 The Liphook pine forest fungal dataset

The counts of fungi listed in Table 1.9 were analysed by PCA on untransformed data, and the results are summarised in Table 6.10. Preliminary inspection of the relative importances of each axis (from the percentage of total variance explained by each one) shows that only the first three axes are likely to be worthy of inspection, since the percentage variance for the fourth axis is below that predicted by the broken stick model (Table 6.8). Further examination of the eigenvector loadings (otherwise called the component loadings or factor loadings) shows that the first axis is weighted towards detection of certain species (especially *Cortinarius cinnamomeus*, *Cortinarius semisanguineus*, *Lactarius rufus* and *Suillus luteus*). An alternative, visual mode of presenting these eigenvector loadings is the bar charts shown in Figures 6.12–6.14. The first axis may be seen to have large positive eigenvector loadings, whereas the only species with a non-trivial negative loading is *Paxillus*

Table 6.10 The outputs from a PCA on the Liphook forest fungal dataset: the analysis included a total of 11 species, so could potentially produce 11 principal axes. Comparison with the broken stick model (Table 6.8) suggested that only the first three axes are worthy of detailed attention.

The eigenvalues and percentage variance				
	axis 1	axis 2	axis 3	axis 4
eigenvalue	3.95	2.89	1.55	0.98
% variance	35.9	26.3	14.1	8.9
cumulative %variance	35.9	62.2	76.3	85.2

Expected values for percentage variance under the broken stick distribution, given 11 variables:

	27.4	18.3	13.8	10.7

From these it can be seen that only the first three axes are likely to be worthy of examination.

	axis 1	axis 2	axis 3
Boletus ferrugineus	0.00	0.88	−0.03
Cortinarius cinnamomeus	0.88	0.15	−0.21
Cortinarius semisanguineus	0.97	−0.10	−0.13
Gomphidius roseus	−0.03	0.94	−0.16
Inocybe lacera	0.52	−0.06	0.71
Laccaria proxima	−0.01	0.13	0.88
Lactarius rufus	0.82	−0.15	−0.31
Paxillus involutus	−0.31	−0.54	0.42
Suillus bovinus	0.04	0.90	0.42
Suillus luteus	0.89	0.17	0.42
Suillus variegatus	0.62	0.09	0.42

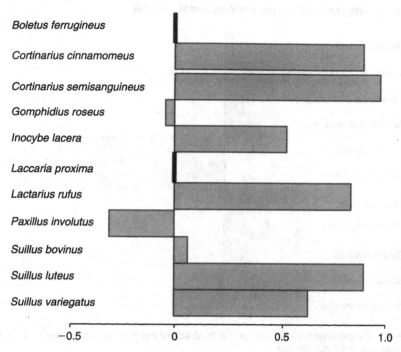

Figure 6.12 The eigenvector loadings for the first axis of the PCA on the Liphook pine forest fungal dataset (Table 6.10).

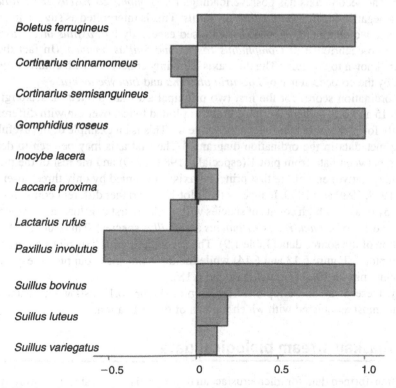

Figure 6.13 The eigenvector loadings for the second axis of the PCA on the Liphook pine forest fungal dataset.

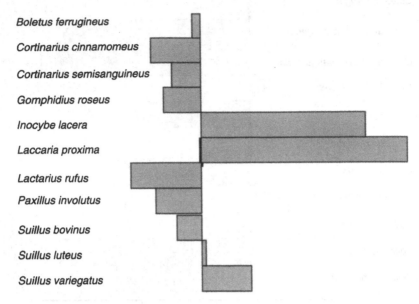

Figure 6.14 The eigenvector loadings for the third axis of the principal components analysis on the Liphook pine forest fungal dataset.

involutus. The second axis has positive loadings for *Gomphidius roseus* and *Suillus bovinus*, and a negative loading for *Paxillus involutus*. This is interpreted as suggesting a gradient between two communities, one characterised especially by *Paxillus involutus* and the other by a co-occurrence of *Gomphidius roseus* and *Suillus bovinus*. (In fact these two species are known to co-fruit.) The third axis is mainly concerned with communities characterised by the co-occurrence of *Laccaria proxima* and *Inocybe lacera*.

These ordination scores for the first two principal axes are plotted on scattergraphs in Figures 6.15 and 6.16, where the same graph is plotted twice, overlain with different symbols, firstly for plot number and secondly for year. (This is an example of the usefulness of including metadata in the ordination diagrams.) The first axis may be seen to detect the difference between data from plot 1 (especially 1988–1990) and the rest of the plots. The right-hand (positive) end of the first principal axis is occupied by only three observations: plot 1 in 1988, 1989 and 1990. It appears that plot 1 had a rather different community, characterised by relatively high counts of species with positive first axis loadings (which we see from Table 6.10 to be *Lactarius*, *Cortinarius* and *Suillus* species). This can be validated by reinspection of the source data (Table 1.9). The second principal axis picks out particularly data from plot 4 (Figure 6.13 and 6.14) while the third axis picks out plot 1, especially the plot 1 community in 1988 (Figures 6.17 and 6.18).

Finally, these data can be replotted as a biplot (Figure 6.19), showing visually which species are most associated with which regions of the ordination.

6.5.2 Alaskan stream biological data

The log-transformed data for microcrustacean densities listed in Table 1.15 were subjected to PCA, and the first two axes of the ordination are presented in Table 6.11. Note that the

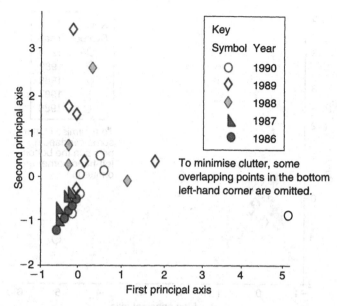

Figure 6.15 The first and second axes of the PCA on the Liphook fungal dataset (untransformed data), overlain with symbols showing year.

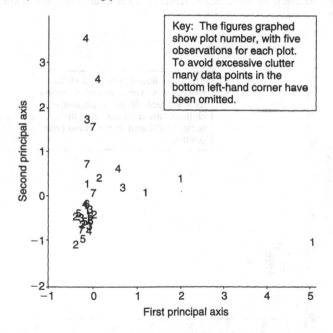

Figure 6.16 The first and second axes of the PCA on the Liphook fungal dataset (untransformed data), overlain with symbols showing plot number.

second axis (16.7% of the variance) explains less than would be predicted under the broken stick model (Table 6.8), suggesting that only the first axis contains interpretable information. The ordination is shown graphically in Figure 6.20 as a biplot, where it may be seen that the first axis of the ordination runs from the assemblage at Berg North (167 years old)

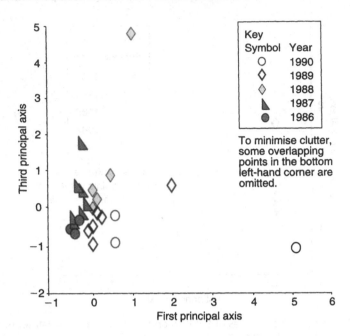

Figure 6.17 The first and third axes of the PCA on the Liphook fungal dataset (untransformed data), overlain with year.

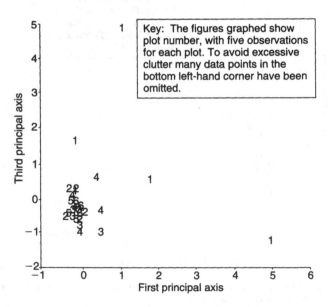

Figure 6.18 The first and third axes of the PCA on the Liphook fungal dataset (untransformed data), overlain with plot number.

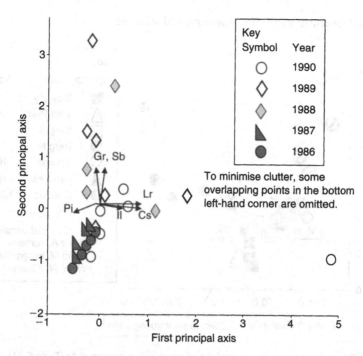

Figure 6.19 The biplot of the first and second axes of the PCA on the Liphook fungal dataset (untransformed data). For clarity some species have been omitted – their loadings are given in Table 6.10.

Table 6.11 PCA of the Alaskan streams biological data after log-transformation. The raw data are given in Table 1.11.

Part 1: The eigenvalues and percentage variance

Axis	1	2
Eigenvalue	3.41	1.86
% variance	31.0	16.9
cumulative		
% variance	31.0	47.9

Part 3: Eigenvector loadings for axes 1 and 2

	axis 1	axis 2
Nitocra hibernica	0.82	0.16
Atheyella illinoisensis	0.86	0.24
Atheyella idahoensis	−0.36	0.71
Maraenobiotus insignipes	−0.11	−0.22
Bryocamptus hiemalis	−0.60	0.42
Bryocamptus zschokkei	−0.45	−0.09
Acanthocyclops vernalis	0.39	−0.39
Alona guttata	0.59	0.40
Graptoleberis sp	0.63	0.00
Chydorus sp	−0.28	0.74
Macrothricidae	0.57	0.41

Part 4: Statistical analyses of ordination scores

The null hypothesis that the first principal axis scores do not differ between streams can be tested by analysis of variance:

ANOVA: $F_{6,28} = 11.9$, $p < 0.001$ both by standard tables and by a Monte Carlo permutation test.

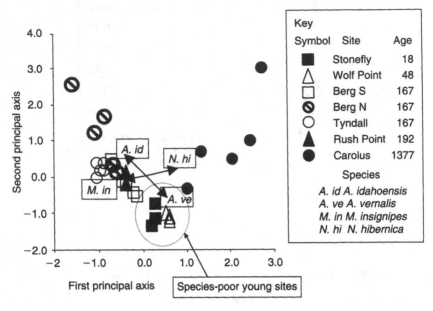

Figure 6.20 The ordination biplot for the PCA of the Alaskan stream data (Table 6.11). For legibility only four species are overlain on the sites ordination.

to the community found in Carolus (1377 years). The species-poor young sites (Stonefly at 18 years and Wolf Point at 48 years) lie in a tight cluster in the middle of the first axis.

This can be interpreted as showing the Berg North and Carolus sites to have the most different communities, which are placed at opposite ends of the first axis. The youngest sites are so species-poor as to approximate to zero, lying 'on the fold' between the two extreme ends.

The ordination diagram does not suggest a simple relationship between either of the first two axes, but it is possible to test the null hypothesis that there is no difference between streams with regards to their axis scores. This was tested by one-way ANOVA, giving an F value of 11.9. Standard tables gave this a significance of less than 1 in 100, which was confirmed by a Monte Carlo test. This shows the sites to differ in their axis scores, but does not in itself explain what the pattern ordination is picking up.

6.5.3 Lichen responses to pollutants

This example concerns the influence of gaseous air pollutants on the lichen *Usnea subfloridana*, and arises from a PhD in lichenology (Ellin, 1989) incorporated in a large outdoor fumigation project (McLeod, 1995) testing the effects of sulphur dioxide (SO_2) and ozone (O_3) on UK forests. The data presented are a small part of a much larger set of analyses, but are included as an interesting example of an ordination generating plausible hypotheses from very weak data.

Table 6.12 part 1 gives mean values of each of six physiological parameters for the seven experimental plots used in the experiment. A fuller analysis would have repeated the ordination for the full dataset rather than just plot means – in fact when this was done the results were very similar. The data consistently showed significant adverse effects of SO_2 on the lichen (as expected for this sensitive species), but few indications of ozone effects.

Table 6.12 PCA ordination of physiological data from a study of the lichen *Usnea subfloridana* exposed to SO_2 and O_3 in a field fumigation system.

Part 1: The raw data

SO_2	O_3	CO_2	Cla	K	Mg	Ca	N	Na
0	0	3.71	106	1.82	337	1.61	14.90	534
H	0	1.84	40	0.82	194	1.43	15.00	400
L	H	3.39	99	0.97	248	1.39	15.00	261
0	H	3.81	212	1.67	331	1.63	15.00	380
H	H	3.70	62	0.72	171	1.20	15.10	410
L	0	3.73	91	1.04	201	1.47	14.80	530
0	0	3.68	99	1.85	300	1.73	14.90	517

SO_2 and O_3 refer to fumigation treatments applied. SO_2 refers to the sulphur dioxide treatment, which was applied at three levels; ambient (0), 14 ppb (L) or 24 ppb (H). O_3 refers to the ozone treatment, applied at ambient (0) or 1.5 × ambient (H). The other variables are means for each fumigation plot of physiological parameters. CO_2 is the rate of photosynthesis, as mg CO_2/g/minute, Cla is chlorophyll A content in µg/g, K is potassium content mg/g, Mg is magnesium content µg/g, Ca is calcium content mg/g, N is nitrogen content mg/g, and Na is sodium content µg/g.

Part 2: The eigenvalues and percentage variance

Axis	1	2
Eigenvalue	3.95	1.57
% variance	56.4	22.4
cumulative		
% variance	56.4	78.9

Part 3: Eigenvector loadings for axes 1 and 2:

	axis 1	axis 2
CO_2	0.56	0.21
Cla	0.69	0.62
K	0.97	0.04
Mg	0.90	0.31
Ca	0.91	−0.10
N	−0.57	0.68
Na	0.51	−0.76

Part 4: Statistical analyses of ordination scores

The hypothesis that the first principal axis corresponds to the influence of SO_2, and the second axis to O_3, can be tested by ANOVA. The design of the experiment was such that these tests are very weak, and no interaction term can safely be tested.

SO_2 ANOVA on first axis scores: $F_{2,4} = 49.6, p < 0.01$.
O_3 ANOVA on second axis scores: $F_{1,5} = 19.6, p < 0.01$

When the data were ordinated the first two axes accounted for 56.4% and 22.4% respectively of the total variance (Table 6.12 part 2). The broken stick approach gives expected values of 40.8% and 24.1% of variance for the first two axes (Table 6.8), on which basis only axis one deserves further examination. The eigenvector loadings (Table 6.12 part 3) show that the first axis weights all variables roughly equally, except for nitrogen content which has a small negative weighting. By contrast, the second axis is dominated by sodium and nitrogen.

The biplot of these two axes is very revealing (Figure 6.21). The first axis appears to correspond to SO_2 effects, with negative scores corresponding to high SO_2 treatments. The

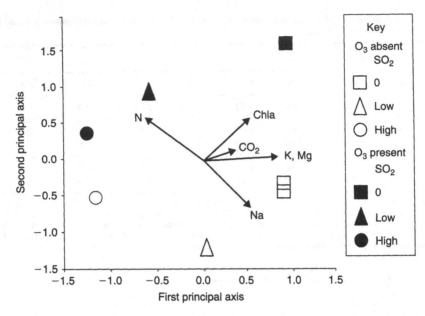

Figure 6.21 The full biplot of the PCA on lichen physiological data presented in Table 6.12.

second axis appears to correspond to O_3 fumigation, with positive scores corresponding to O_3 treated plots. The separation of the plots according to their treatments is effectively perfect!

These hypotheses can be tested by ANOVA, although the unbalanced design of the experiment makes the ANOVA models used rather weak. These analyses are summarised in Table 6.12 part 4, and confirm the patterns observed to be significant.

An inspection of the arrows on the biplot shows clearly that all the cations (Ca, K, Mg, Na) have responded in the same way, pointing towards the right-hand end of the first principal axis (meaning that the highest concentrations of cations occur in the zero SO_2 treatments). These elements had a small projection on the second axis – in other words the first axis (SO_2 treatments) reduced concentrations of all cations, while the second axis (ozone) had a negligible effect on cation content. The arrow pointing to nitrogen behaves very differently, pointing to the high SO_2 and high ozone corner of the ordination. This element was actually increased both by SO_2 and by ozone (although inspection of the raw data shows these effects to be weak). As a technical detail, these observations may be explained by two different chemical processes. The increase of nitrogen with ozone treatment may have been because of contamination of the ozone by dinitrogen pentoxide (Brown and Roberts, 1988). The increase of nitrogen under the high SO_2 treatment is explicable by the co-deposition of ammonia with SO_2 (McLeod et al., 1995).

This example illustrate the power of PCA, and the difficulties inherent in using the technique. It found a pair of contrasting treatment effects in data that fitted with what might have been expected, and produced a near-perfect ordination of the experimental treatments. The eigenvector loadings (shown as the direction of the arrows on the biplot) fitted accurately with what is known about the responses of each variable to the treatments imposed. However the second axis should technically have been ignored as random noise, and even with the benefit of hindsight it is hard to find the ozone effect in the raw data.

7

Correspondence analysis, simple and detrended

7.1 General introduction

The aim of this chapter is to introduce two closely related ordination techniques, correspondence analysis (CA) and detrended correspondence analysis (DCA). Before exploring their application it would be useful briefly to review the history and relationship between these two techniques.

Correspondence analysis also goes under the name of reciprocal averaging (RA), and at its core involves repeatedly calculating weighted averages (iteratively weighting sites by dependent variables, and then dependent variables by sites – details are given below). 'Reciprocal averaging' is certainly the better name since it describes the mathematical basis of the ordination, but 'correspondence analysis' is the term that has gained widest usage and has been incorporated into the names of derived techniques such as detrended correspondence analysis and canonical correspondence analysis. Two other names, once applied to this technique but which have now vanished, are reciprocal ordering (Orloci, 1978) and dual scaling (Nishisato, 1980).

Perhaps because of the relatively simple structure of the calculations needed, CA has the unusual distinction of having been rediscovered far more often than any other ordination technique (de Leeuw, 1983). Correspondence analysis dates back to the work of Hirschfeld (1935) and Fisher (1940), but remained obscure until reintroduced to ecology by Mark Hill (Hill, 1973b; 1974). Hill also highlighted certain recurring problems when CA is applied to ecological data, specifically arch effect; the production of curved trends in data following a smooth gradient, and went on to produce a FORTRAN package which implemented certain modifications to the basic CA algorithm in order to solve these problems. The package was called DECORANA (Hill, 1979a; Hill and Gauch, 1980) an acronym which tends to be used interchangeably with the correct name for the underlying algorithm of detrended correspondence analysis. DECORANA has remained a widely available package within the ecological research community to this day. In consequence ecologists tend not to use simple CA but rely instead on DCA. Outside of ecology, CA has remained popular as a basic ordination technique only in France, due in part to the efforts of Benzécri to promote this ordination as an analytical tool (Benzécri, 1992). This is a pity, since CA is comparable to PCA in its utility for data interpretation, and has the virtue that sites and dependent variables are ordinated at the same time, facilitating the production of biplots.

Detrended correspondence analysis (DCA) is an ordination technique based on CA but which was designed specifically to assist in the interpretation of ecological data. The modifications needed to 'improve' the basic CA algorithm are rather crude, and based on an empirical desire to reshape data closer to the models visualised by ecologists. Despite this, DCA has consistently performed so well when applied to ecological data (simulated and real) that it has remained one of the most widely used ordination techniques within population ecology, and is a standard part of the ecologist's analytical toolkit (at least for Anglophone European and American countries, though less so in Australia). Outside ecology the technique is virtually unknown; the author has encountered university statisticians who knew nothing about DCA. This also is a pity since it has potential for application to non-ecological data, specifically in situations where measurements are taken along the length of a long steady gradient.

7.2 Correspondence analysis

7.2.1 Weighted means and related matters

The idea underlying correspondence is the weighted mean. This is a simple extension of the school-level concept of the average of a set of numbers; add them up and divide by the number of observations. Statisticians do exactly the same thing, but call the result a 'mean'.

A weighted mean results when some of the numbers in the data are repeated. To take a simple example, the unweighted mean of 1, 2, 3, 4, 5 is 3 ($1 + 2 + 3 + 4 + 5 = 15$; $15/5 = 3$). If these were scores from a trial, and the score of 1 occurred 10 times while the other scores occurred once, a new mean would be calculated in which the 1 is given 10 times the weight of the other numbers:

$$\text{Weighted mean} = \frac{(10 \times 1 + 2 + 3 + 4 + 5)}{10 + 1 + 1 + 1 + 1} = \frac{24}{14} = 1.7143\ldots$$

It will be seen that the new mean has been reduced by the increased importance (hence 'weight') given to the number 1.

We will consider a simple model dataset, given in Table 7.1, describing a time series in which three species in turn dominate their community. We can now use the concept of the weighted mean to calculate a 'mean year' for each of the three species. This is done by weighting each year by the frequency of a chosen species, then calculating the weighted mean of the years.

Formally this would be written as:

$$\text{Mean year} = \frac{\sum(\text{year} \times \text{species count})}{\sum \text{species count}} \tag{7.1}$$

Thus for species A, the mean year is found as follows:

$$\text{Mean year} = \frac{1 \times 100 + 2 \times 90 + 3 \times 80 + 4 \times 60 + 5 \times 50 + 6 \times 40 + 7 \times 20 + 8 \times 5}{100 + 90 + 80 + 60 + 50 + 40 + 20 + 5}$$

$$= 3.21 \text{ years}$$

Table 7.1 The algorithm used in correspondence analysis, based on a model dataset describing a succession involving three species, A, B, and C (respectively early, mid- and late successional). The calculations performed here are explained in the text.

Year	Counts of species			Y_1	Y_2	Y_{15}
	A	B	C			
1	100	0	0	0.0	0.0	0.0
2	90	10	0	5.0	4.2	3.9
3	80	20	5	14.3	12.7	12.2
4	60	35	10	26.2	23.4	22.5
5	50	50	20	37.5	34.0	32.9
6	40	60	30	46.2	42.3	41.1
7	20	30	40	61.1	58.3	57.5
8	5	20	60	82.4	80.4	79.8
9	0	10	75	94.9	93.1	92.8
10	0	0	90	100.0	100.0	100.0

Weightings for each species on successive iterations.

Species	A	B	C
Iteration 0 (start):	0.0	50.0	100.0
S_{1a}	19.1	43.9	78.5
S_{1b}	0.0	41.7	100.0
...			
S_{15}	0.0	39.0	100.0

For species B and C the mean year values are 5.49 and 8.09 years, respectively. Note that these values confirm the assertion made earlier that species A is early successional, species B is mid-successional and species C is late successional.

This calculation is of some use and interest in itself, and leads on to the next key step in the algorithm.

Having defined a 'mean year' weighting for each species by use of a weighted mean, it is also possible to define a 'mean species' (or more accurately a mean species weight) for each year the same way. This is better termed the year weighting, and would be calculated as:

$$\text{Year weighting} = \frac{\sum(\text{species weight} \times \text{species count})}{\sum \text{species count}} \quad (7.2)$$

7.2.2 The CA algorithm

We can now link together the approaches defined in equations 7.1 and 7.2 to understand the CA algorithm.

The sequence starts by giving an arbitrary set of weightings to each species, subject to the constraint that the species weightings run evenly from a minimum value of 0.0 to some defined maximum value, which is usually taken to be 100. This is achieved in Table 7.1 by giving the three species weightings of 0, 50 and 100. More generally, if there is a total of S species, the initial weightings for the nth species are usually set at $100 \times (n - 1)/(S - 1)$, although in fact different values will not affect the final result.

These initial species weightings are shown at the bottom of Table 7.1 on the line labelled 'Iteration 0'. We can now use these to calculate a weighting for each of the years using equation 7.2 as follows:

$$\text{Year 1 weighting} = \frac{0 \times 100 + 50 \times 0 + 100 \times 0}{100 + 0 + 0} = 0.0$$

$$\text{Year 2 weighting} = \frac{0 \times 90 + 50 \times 10 + 100 \times 0}{90 + 10 + 0} = 5.0$$

The weightings for the other years are calculated in the same way, and are given in Table 7.1 under the column Y_1. These year weightings can be used to calculate new species weightings, using equation 7.1. The new weighting for species A is:

$$\frac{(0 \times 100 + 5 \times 90 + 14.3 \times 80 + 26.2 \times 60 + 37.5 \times 50 + 46.2 \times 40)}{(100 + 90 + 80 + 60 + 50 + 40 + 20 + 5)}$$

$$+ \frac{(61.1 \times 20 + 82.4 \times 5 + 0 \times 94.9 + 0 \times 100)}{(100 + 90 + 80 + 60 + 50 + 40 + 20 + 5)} = 19.1$$

The weightings for species B and C are worked out the same way, and given in Table 7.1 as iteration S_{1a}.

The new species weightings may be used directly to calculate a second set of year weightings, thence a third set of year weightings, etc. If this procedure is followed directly it will be found that the weightings all converge onto one constant value, which is unhelpful. Instead it is necessary to rescale the species weightings back into the range 0–100. If the minimum species loading is MIN and the maximum is MAX, the transformation between original species weightings (S_{1a}) and rescaled values (S_{1b}) is given by the formula

$$S_{1b} = \frac{100 \times (S_{1a} - \text{MIN})}{\text{MAX} - \text{MIN}} \tag{7.3}$$

The rescaled species weightings are given in Table 7.1 on the line S_{1b}. This completes the first cycle of the CA algorithm. (The rescaling procedure is often skipped by textbooks when explaining CA, leading to unexpected and unhelpful results for users trying to write their own programs to execute this procedure.)

The process is now repeated a second time. The weighting for each year is recalculated using the new, rescaled weightings for the species. These second cycle year weightings are given in Table 7.1 under the column Y_2. Then the weightings for each species are recalculated using the new year weightings, and rescaled. This exercise is repeated until a stable pattern emerges (typically 10–20 iterations will suffice). The example data stabilise after 15 iterations, and the values are given in Table 7.1 as Y_{15} (year weightings) and S_{15} (species weightings). It should be noted that these final values are independent of the actual choice of weightings given to each species at the beginning of the operation, and give the first CA ordination axis for the rows (the year weightings) and the columns (the species weightings).

A second ordination axis may be derived in the same way, by repeating the weighted mean calculations but including a term that removes the influence of the first ordination axis.

Although tedious for humans, this algorithm is straightforward and efficient for computers, with the amount of work rising linearly with the size of the dataset. When software packages implement CA they generally use this iterative algorithm for its computational efficiency and stability. However there are other ways into CA whose connection with weighted means is far from obvious. For one, it can be shown that this procedure can be rewritten as an eigenvector problem, with the ordination being given by the eigenvectors of the matrix of χ^2 values that would be used to test the hypothesis that species are distributed randomly with respect to each other (Chardy, Glemarec and Laurec, 1976). CA is also equivalent to simultaneously assigning weightings to rows and separate weightings to columns within the same dataset, so that the correlation between the resulting pair of variables is maximised.

It should be noticed that the species and sites ordinations are run simultaneously, and hence may be overlain directly onto the same ordination diagram.

7.2.3 Limitations of correspondence analysis

CA ordination produces a space which reflects the χ^2 values between species; in other words species that are negatively associated are placed far apart. In this respect CA matches PCA, although the exact pattern generated will differ between the two techniques. Consequently CA suffers from the arch effect (Section 7.3) although not quite as severely as does PCA.

Although CA ordinations produce eigenvalues, these are less readily interpreted than in PCA. They tend not to add to 1.0, but instead to a value known as the inertia of the dataset. Each progressive eigenvalue decreases in size, as in PCA, and dividing each eigenvalue by the total inertia of the data shows what percentage of the total variation is explained by each axis, but unlike PCA there are no accepted stopping rules. Most packages that run CA provide the first three axes, and leave their interpretation up to the researcher. A rather harsh rule of thumb suggested by McCune and Mefford (1999) is that CA should not be used if the data contain more than one axis of variation.

There are other, more serious problems with CA as an ordination technique. The use of the χ^2 statistic results in the ordination space being distorted, with points clumped together more tightly near the ends of ordination axes than in the middle (Hill, 1979a). In addition it is overly sensitive to the occurrence of scarce species in the data (because they tend to generate high χ^2 values). The sensitivity to rare species is of sufficient concern that the standard implementation of both CA and its derivation DCA contain an option to down-weight rare species.

Perhaps because of these reservations CA ordination is used rather less than PCA, being cited in approximately 60% as many publications as PCA for the period 1990–1996. It is included here both as a useful technique in its own right, and because it forms the basis of three more widely used analytical techniques: DCA (below), the TWINSPAN classification package (Chapter 8), and canonical correspondence analysis (Chapter 9).

7.2.4 Worked example: CA ordination of the Liphook pine forest fungal dataset

The counts of fungi listed in Table 1.9 were analysed by CA on untransformed data, and the results are summarised in Table 7.2. The eigenvalues show that the first three axes accounted

Table 7.2 The results of running CA on Liphook pine forest fungal data.

Part 1: The sites ordination

Year	Plot	CA axis 1	CA axis 2	CA axis 3	Year	Plot	CA axis 1	CA axis 2	CA axis 3
86	1	195.04	7.56	−43.18	88	5	160.89	2.10	3.70
86	2	211.48	10.34	165.65	88	6	113.82	−1.15	−30.88
86	3	246.20	16.22	606.63	88	7	−3.77	−24.38	−10.29
86	4	198.24	8.10	−2.50	89	1	−111.66	130.93	−0.10
86	5	254.26	17.59	708.99	89	2	−99.49	6.55	4.25
86	6	207.19	9.62	111.18	89	3	−105.75	−44.87	6.34
86	7	223.23	12.34	314.84	89	4	−98.12	−43.07	3.33
87	1	192.46	7.88	−45.19	89	5	−67.54	−26.99	31.94
87	2	167.40	1.83	−2.43	89	6	−93.35	28.77	4.31
87	3	148.18	−1.83	90.88	89	7	−107.32	−36.18	5.10
87	4	106.59	−10.71	−23.00	90	1	−116.84	479.29	5.25
87	5	213.36	10.66	189.58	90	2	−81.34	−0.55	8.41
87	6	195.03	7.55	−32.52	90	3	−90.11	−21.01	6.98
87	7	176.94	4.44	45.76	90	4	−81.30	−11.38	1.88
88	1	116.98	28.66	−34.86	90	5	−26.68	−6.73	65.22
88	2	51.22	−9.23	−18.75	90	6	−45.54	1.22	32.65
88	3	−7.05	−28.05	−11.97	90	7	−101.75	−26.13	7.53
88	4	14.25	−20.14	−15.97					

Part 2: The species ordination

Species	axis1	axis 2	axis 3
B. ferrug	−30.01	−32.90	−32.29
C. semisa	−178.63	864.44	10.56
G. roseus	−102.41	−80.02	−1.47
I. lacera	74.59	454.98	−123.68
L. proxim	194.78	7.52	−46.46
L. rufus	−174.60	696.97	74.12
P. involu	267.48	19.82	876.87
S. bovinu	−109.13	−41.28	6.70
S. luteus	−127.80	516.96	−15.46
S. varieg	−126.42	352.88	−10.93

(For full species names see Table 1.9).

Part 3: The eigenvalues
Total inertia of data: 1.16

Axis	eigenvalue	% variance
1	0.545	47.0
2	0.326	28.1
3	0.177	10.1

for 47, 28.1 and 10.1% of the variance, respectively, although no guidelines exist for formally interpreting these values.

When the first two axes of the species ordination are graphed (Figure 7.1), the points form a shape like a capital 'L', with early successional species (*Paxillus involutus, Laccaria proxima*) in the bottom right-hand corner while later successional species (*Cortinarius semisanguineus, Lactarius rufus* and *Suillus luteus*) occupy the upper left. *Gomphidius roseus* and *Suillus bovinus* are closely adjacent in the middle of the graph. There is one odd placing, *Inocybe lacera*, which lies outside the 'L' high up in the middle of the ordination.

The sites ordination (Figure 7.2) shows a similar L-shaped pattern of points to the species ordination, with data from the early years in the bottom right and the later years in the bottom left (hence age is spread out along the first axis). The second axis, the upper arm of the 'L', are 1989 and 1990 data from plot 1.

These two graphs can be overlain to produce a full CA ordination biplot (Figure 7.3), which may be used as an illustrative representation of the whole dataset. It shows a progression

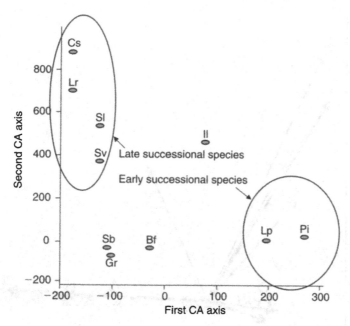

Figure 7.1 The correspondence analysis of the Liphook pine forest fungal dataset, showing only the species ordination. Abbreviations are listed in Table 1.9.

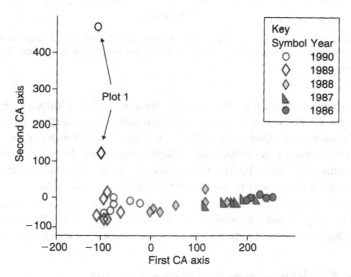

Figure 7.2 Correspondence analysis of the Liphook pine forest dataset, showing only the site ordination (overlain with year).

along the first axis (going from right to left) with the youngest communities dominated by *Laccaria proxima* and *Paxillus involutus* in the bottom right, moving to the later communities dominated by *Suillus bovinus*. Plot 1 veers off on a different course in 1989 and 1990, towards a community dominated by *Cortinarius, Lactarius* and *Suillus*.

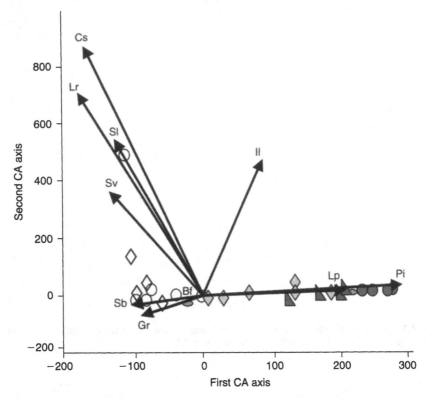

Figure 7.3 The correspondence analysis of the Liphook pine forest fungal dataset, showing both the species and site ordinations overlain. Abbreviations are as in Figures 7.1 and 7.2; species are shown by arrows.

It is instructive to compare the CA ordination with the biplot produced by PCA on these same data (Figure 6.19). There is little superficial similarity, but in fact (as would be expected) both show the same underlying pattern. Both ordinations detect a major axis of variation which runs from the early years (1986, 1987) where *Laccaria* and *Paxillus* dominate, to a following community dominated by *Suillus bovinus*. They also identify a second axis which detects an anomalous community balance in plot 1 in 1989 and 1990. However, CA detects the successional trend along its first axis, and separates the years quite accurately except for 1989/1990, relegating the plot 1 anomaly to its second axis. By contrast, PCA detects the plot 1 outliers as the biggest source of variation, placing successional change along its second axis.

7.2.5 The Alaskan microcrustacean dataset

The data from Table 1.5 were ordinated by CA. The ordination scores are given in Table 7.3, while a slightly simplified version of the ordination diagram is shown in Figure 7.4. The sites ordination may be seen to separate off the data from the youngest site (Stonefly, at 48 years) as an outlier on the second axis, while the remainder of the sites lie along the first axis, clearly separating out the oldest site (Carolus, 1377 years) at the right-hand end of the first axis. The species ordination matches this pattern, with the species (*M. insignipes*) found at Stonefly

Table 7.3 The CA ordination of the Alaskan stream dataset (source data in Table 1.14). This ordination is graphed in Figure 7.4.

	CA axis		
	1	2	3
% variance	27.1	24.6	16.1
Cumulative %	27.1	51.7	67.8
The Sites ordination			
stonefly	−376.11334	973.36700	26.17672
stonefly	−376.11334	973.36700	26.17672
wolf pt	146.86086	−0.19624	−197.63560
wolf pt	146.86086	−0.19624	−197.63562
wolf pt	146.86087	−0.19624	−197.63560
wolf pt	146.86086	−0.19624	−197.63562
wolf pt	146.86086	−0.19624	−197.63562
tyndall	−147.60596	106.41737	−22.51982
tyndall	−163.28726	−77.51281	4.91455
tyndall	−94.57098	−59.85105	−43.12987
tyndall	−163.27798	−77.50758	4.89109
tyndall	−165.02336	−78.49131	9.30032
berg n	−165.22397	−78.60437	9.80702
berg n	−206.82051	−124.63148	165.12898
berg n	−183.36366	−100.84338	79.94843
berg n	−182.09801	−88.11488	52.43434
berg n	−198.72583	−112.89433	128.57504
berg s	−98.26357	−74.21804	3.20000
berg s	−148.34991	−69.09386	−32.82029
berg s	−70.84834	−54.30967	−56.40331
berg s	−23.97962	−40.06776	−102.25587
berg s	−75.23978	−55.99677	−50.00542
rush pt	−177.86058	−102.41098	75.49634
rush pt	146.86086	−0.19624	−197.63562
rush pt	−182.09799	−88.11488	52.43434
rush pt	−45.30341	−48.13351	−71.94240
rush pt	−182.09801	−88.11488	52.43434
carolus	479.68524	78.04498	219.69421
carolus	333.04700	43.24961	22.28313
carolus	372.60107	52.26548	59.08732
carolus	402.86749	59.84422	117.21169
carolus	352.15027	47.47250	33.97442
The Species ordination			
lognito	461.16086	73.14450	173.60825
logatill	451.22910	70.45534	144.52942
logatid	−207.37126	−135.72812	183.81296
logmarin	−376.11334	973.36700	26.17672
logbryhi	−182.09799	−88.11488	52.43434
logbryzs	−148.34991	−69.09386	−32.82029
logacver	146.86086	−0.19624	−197.63562
logalong	522.94739	89.48962	327.32428
loggrapt	433.39685	65.63747	93.39194
logchydo	−242.51418	−156.31549	294.74625
logmacro	481.82458	78.76688	235.25519

being placed in the same outlier position as Stonefly itself, while a suite of five species found in the oldest site are almost superimposed on each other, pointing towards the right-hand end of the first axis. This is a different pattern to that produced by PCA of the same data, which placed the youngest sites in the centre of the first ordination axis (Figure 6.20).

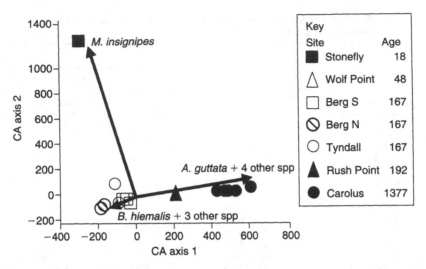

Figure 7.4 CA ordination of the Alaskan streams dataset. (Some observations and species have been removed to reduce clutter.)

7.3 Detrended correspondence analysis

Unlike PCA or CA, DCA is not a mathematically 'natural' technique. The results that come out are not an accurate reflection of the shape of the data supplied. Its operation cannot, even in principle, be reduced to a simple set of matrix equations. Instead the data supplied are reshaped according to a defined set of rules. These reshaping rules are applied iteratively until a specified level of accuracy is achieved, so that the same dataset can generate different outputs depending on the settings used by the package. Despite this, DCA is one of the most widely used ordination techniques in ecology because it was written with the intention of overcoming a specific weakness in standard ordination techniques. In this chapter, as in most other literature about DCA, it is assumed that data are ecological, with columns referring to species and rows to samples. In principle there is no reason why the analysis should not be used on chemical, geophysical or other forms of data although variables with negative values must be excluded.

The weakness in question has been alluded to earlier (Chapters 5 and 6), and is known as the arch effect (also the horseshoe effect). It arises whenever PCA or other distance-conserving ordination techniques are applied to data which follow a continuous gradient, along which there is a progressive turnover of dominant variables. The classical example of this is an ecological succession, in which one community of organisms is gradually replaced by another, which in turn changes again. Gradients in space (transects) can generate the same problem.

When such data are ordinated by a distance-conserving technique and the first two axes are plotted against each other it will be found that successive data points form a curve which is typically shaped roughly like a letter U. (In PCA the ends bend in towards each other like a horseshoe. In other ordination techniques the problem is marginally less severe, and the resulting shape is more like a smooth arch.) This same arch distortion is readily seen in simple ordinations of the Liphook pine forest dataset (Figures 5.4, 6.19, 7.3) which is primarily a dataset describing an ecological succession.

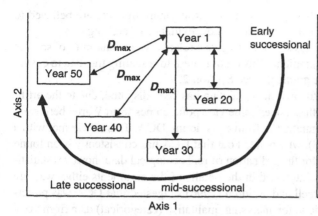

Axis 2 / Axis 1

Year 1 — Early successional
D_{max}
Year 50 — D_{max}
D_{max}
Year 20
Year 40
Year 30
Late successional   mid-successional

The ordination operates in a space where closeness between two observations indicates that they contain similar data. The greatest possible separation, indicating no overlap, is given by the distance D_{max}. This distance must separate the observation for year 1 and all observations after year 20. The only way to ensure this is for the succession to be portrayed as a curve. Compare this figure with Figure 5.4

Figure 7.5 A diagrammatic representation of the arch effect when ordinating data from a steady gradient. Here it is assumed that there is little or no similarity between the data for year 1 and for years 30 upwards.

As the succession proceeds (or the gradient lengthens) the horseshoe distortion becomes more intense, and in extreme cases the distortion becomes so great as to be the largest source of variation in the data – so becomes the first ordination axis. This leads to the paradox that the overriding trend in the data, the successional change, may be found on the second ordination axis, with the first axis representing the horseshoe effect – a mathematical artefact.

This effect can lead to misinterpretation of data. If analysis were to be confined to the first axis, it would be possible to conclude that no succession were occurring in a highly successional dataset. Since diversity tends to peak mid-succession (Brown and Southwood, 1987) the first axis score may correlate significantly with indices of diversity, leading to the spurious conclusion that the first ordination axis represents community diversity.

Early observations of the horseshoe effect led to speculation that it reflected a deep result in community ecology. It doesn't. It reflects the simple mathematical fact that basic ordinations such as PCA are a shape-conserving rotation of points in some data space. The distance between communities in data space is a reflection of how similar their community compositions are. For any set of N species, when represented in N-dimensional data space, there will be some distance D_{max} which corresponds to 'No species in common at all'. The precise value of this distance is immaterial.

Consider an early successional community (pioneer species, or those at one extreme end of a spatial gradient). It will have a number of characteristic species, probably found nowhere else in the dataset. It may share no species in common with a mid-successional community, so the distance between the two communities in N-dimensional data space must be D_{max}. Similarly, the distance between the pioneer community and the late successional community must also be D_{max}. This relationship is faithfully retained after ordination, which will show the mid-successional and late successional communities at the same distance from the pioneer community. This can only be achieved by projecting the succession into a curve – hence the horseshoe. This explanation is shown graphically in Figure 7.5.

Although the arch effect is not actually an error, it is a pitfall for the unwary when interpreting ordinations. DCA ordination is designed specifically to deal with this common problem by removing the arch from the data, using a rather crude method explained below.

Consequently one would consider using DCA in situations where the data are believed to lie along a gradient, or where ordination diagrams show evidence of arching.

DCA also rescales its ordination axes in order to give a relatively constant rates of species turnover along the environmental gradient. This is equivalent to rescaling the data to ensure a constant beta diversity along the gradient – see Section 2.1.

The wisdom of applying DCA to environmental data has been disputed, due to the artificial nature of the ordination algorithm. An extreme viewpoint comes from Wartenberg *et al.* (1987), who tried to dissuade potential users from considering DCA. This article met with a firm refutation by Peet *et al.* (1988), who pointed out that DCA has consistently been found to be superior to other techniques for the ordination of real ecological data, however statistically inelegant it may be. Readers interested in the details of the arguments either way are referred to the original papers. Parnell and Waldren (1996) suggested that DCA was preferable to any other ordination technique for analysing qualitative (categorical) data from taxonomic studies. They did however make the curious observation that the numerical values allocated to bistate data affected the resulting ordination; thus coding bistate data as 1 or 0 gave different results to coding it as 1 or 100. More seriously, Oksanen and Minchen (1997) reported the existence of a minor bug (programming fault) in the original DECORANA code, which caused the ordination to give order-sensitive output for certain datasets. The order in which data points are entered into an indirect ordination should not affect the resulting scores, but this criterion was not true for some datasets under old versions of DECORANA. The bug concerns a stopping condition in a smoothing routine and has been fixed in recent releases of the software, but the researcher who plans to include DCA output in a write-up should show awareness of this problem and specify which software was used. (In the case of this book the software was PC-ORD 1999, which includes the updated version of the code.)

7.3.1 The mechanics of DCA

DCA ordination is a two-stage process. First the data are ordinated by CA, as described above. The algorithm now sorts out the arch effect by a brute-force approach. The ordination scores for the first two axes are assumed to be arched, and this arch is divided up into a number of short segments (typically 26). Each segment is now transformed so that the mean second axis score is zero (see Figure 7.6). This process is repeated until a desired level of accuracy is reached; typically three iterations are adequate. The end result is the guaranteed removal of the arch distortion, leaving an ordination in which a steady gradient is represented as a steady progression along the first axis. Fortunately, most users never need to worry about the precise details of the DCA algorithm, since the package uses sensible default values and has proved to be a robust analytical tool.

Despite apparently destroying the second ordination axis by this detrending operation, it is possible to extract second and higher axes in DECORANA. The CA ordination of the data is simply repeated but including terms which remove the effect of the first DCA axis. The detrending algorithm is again invoked to remove distortion from higher axes. Third and higher axes may also be created.

After detrending the data, DCA corrects another fault of CA, namely its tendency to space data more widely in the middle of the axis than at the ends. This is done by further rescaling of the axis to ensure that variance is evenly distributed.

As with other indirect ordinations, DCA axes are mutually orthogonal, and become progressively less important (in terms of variance explained – not necessarily in terms of their

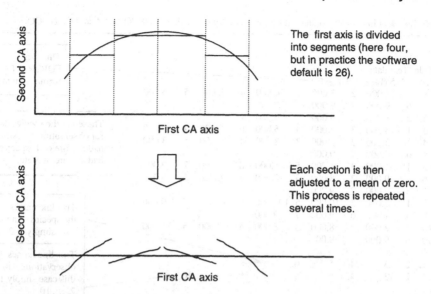

Figure 7.6 The detrending algorithm used by DECORANA.

usefulness to the researcher). There are no firm stopping rules for DCA, and the software default is to produce the first three axes. As with CA, the importance of each axis is indicated by its eigenvalue but interpreting importance of the eigenvalues is a rather subjective operation. (The eigenvalues sum to approximately, but not exactly, 1.0. Due to the iterative operations inherent in obtaining DCA scores it is not possible to give a precise interpretation of this sum of eigenvalues.) The species and sites ordinations are produced simultaneously and may be overlain on a biplot.

7.3.2 DCA software

One of the aims of this book is to be software-independent, due to the rapid development of software. Consequently the specifics of input and output formats have been entirely ignored. This rule will be broken for DCA, since all commercial implementations of the algorithm that the author is aware of call up the same source code. There are now several different 'front-ends' to enter data into DECORANA, but the procedure followed will be the same and the output will always be in much the same format. Consequently it is worth examining the details of this package. This contrasts with simple multivariate techniques such as PCA and MLR which are available in a wide variety of packages with very different input and output formats.

Input format

DECORANA was written in FORTRAN, and the original version required the user to have some knowledge of FORTRAN format statements to enter data. These format statements are used to tell the package exactly which columns in a datafile to read, and what these columns should contain. They are a hang-over from the early days of computing, and nowadays relatively few computer users are competent to define them correctly. (The author has painful memories of trying to persuade FORTRAN compilers to input even trivial data files due to problems with format statements.)

Table 7.4 Data in Cornell Condensed Format, used by the packages DECORANA and CANOCO.

	Title: Test data									
	(I4, 1X, 5 (I4, F10.3))									
1	1	1.000	2	3.000	3	4.000	4	2.000	5	6.000
1	6	9.000	7	9.000						
2	2	1.000	6	9.000	7	9.000				
3	1	3.600	3	24.000	4	5.800	6	9.000	7	9.000
4	1	3.000	2	3.000	3	3.000	4	3.000	5	3.000
4	6	9.000	7	9.000						
5	1	1.000	3	1.000	5	1.000	6	9.000	7	9.000
6	1	9.000	2	8.000	3	7.000	4	6.000	5	5.000
6	6	9.000	7	9.000						
7	1	1.000	3	10.000	5	100.000	6	9.000	7	9.000
9	5	1.000	6	9.000	7	9.000				
10	1	8.000	2	8.000	3	8.000	4	8.000	5	8.000
10	6	9.000	7	9.000						
0										
1.	2.	3.	4.	5.	6.	7.				
1.	2.	3.	4.	5.	6.	7.	8.	9.	10.	

Annotations:
- This is a FORTRAN format statement
- These are the species data: for observation 1, species 1 had a score of 1, species 2 had a score of 3 etc.
- This line names the species – in this case simply 1, 2, …, 7
- This line names the observations – in this case simply 1, 2, …, 10

This is equivalent to the following data matrix Species.

	1	2	3	4	5	6	7
Obs 1:	1	3	4	2	6	9	9
Obs 2:	0	1	0	0	0	9	9
Obs 3:	3.6	0	24	5.8	0	9	9
Obs 4:	3	3	3	3	3	9	9
Obs 5:	1	0	1	0	1	9	9
Obs 6:	9	8	7	6	5	9	9
Obs 7:	1	0	10	0	100	9	9
Obs 8:	0	0	0	0	0	0	0
Obs 9:	0	0	0	0	1	9	9
Obs 10:	8	8	8	8	8	9	9

The same data in the matrix format required by the original Cornell conversion program.

Title: Test data

7	10	4				
(7F4.0)						
1	3	4	2	6	9	9
0	1	0	0	0	9	9
3.6	0	24	5.8	0	9	9
3	3	3	3	3	9	9
1	0	1	0	1	9	9
9	8	7	6	5	9	9
1	0	10	0	100	9	9
0	0	0	0	0	0	0
0	0	0	0	1	9	9
8	8	8	8	8	9	9

In addition, the program requires the data to be in a special format called Cornell Condensed Format. An example of this format is given in Table 7.4. It is an extremely efficient way of presenting data containing many zeros (as is typical for ecological data), but is not immediately accessible to most humans. Consequently the original version of DECORANA (which can still be found) came with a small application that converted a rectangular matrix into Cornell Condensed Format. This application took a rectangular matrix of data in the usual species × sites format, but required three additional items of information (see example in Table 7.2). The first line of the file needed to be a title for the analysis. This was followed by two lines of formatting information. These described the number of variables and of observations, and had to be located in exactly the correct columns. The author wasted an entire day trying to get this application to work, before realising that it needed an extra space at the start of the first format statement!

As recently as 1993, DECORANA still needed data in Cornell Condensed Format, but happily recent versions now accept data in a spreadsheet format. Remarkably, the original software did not hold data in two-dimensional matrices, but relied entirely in the Cornell format. This allows it to handle large, sparse matrices (i.e. those containing many rare species) without filling the computer's memory with zeros. (Computer memory space was limitingly expensive in the 1970s.) This remarkable computational efficiency is achieved at the cost of rather impenetrable FORTRAN coding and a non-standard data format, which explains in part why conventional statistical packages have not rewritten the algorithm and included it among their standard statistical functions.

Output format

Although the DECORANA input format has been rendered more user-friendly in modern implementations, the output format has remained standard.

Consequently, it is worth becoming acquainted with this output (unlike PCA, which is implemented by numerous different packages with very differing output formats). An example is shown in Table 7.5, with actual output in bold and explanatory comments in normal text.

The output starts with some information about the parameters used. These are: transformations, downweighting, rescaling, number of segments to divide the arch into, and a sensitivity parameter. These terms will be explained in turn.

Downweighting of rare species CA and DCA are overly sensitive to the presence of rare species, so a downweighting option is available to selectively decrease the importance of scarce species. In the example shown the option was not used.

Rescaling This is the process which turns CA into DCA; not surprisingly it must be selected in order to run DCA.

The other two parameters are defaults, and rarely altered by users.

This is followed by information about the iterations needed and the length of segments used – which are rarely examined by users beyond a cursory glance.

Finally come tables of weightings, which is the section of genuine importance to most users. There are two such tables, one for the species weighting and one for the observations' weightings. Both tables take the same format, and for the sake of simplicity only the species weightings are shown in Table 7.5.

DECORANA always produces a constant number of ordination axes, unlike PCA, where packages generally allow a choice in this matter. (Some versions give three axes, others four.) The examples shown here came from a version of the program generating three axes. These are shown under the headings of AX1, AX2 and AX3. As with PCA, AX1 is the most important axis hence has the largest eigenvalue.

Each eigenvalue is given twice, once at the end of the information about the residuals and again in the table of weightings. The second time it is abbreviated to EIG =; in the example of Table 7.5 the first eigenvalue was 0.545.

It is important to notice that DECORANA eigenvalues sum to (approximately) 1.0, rather than the number of variables as is the case for PCA. (The actual eigenvalues supplied usually sum to rather less than 1.0, due to unseen higher axes.) Thus an eigenvalue of 0.54 corresponds to approximately 54% of the variance. The first eigenvalue is the same as that produced by CA alone, but subsequent eigenvalues will differ. Typically the second eigenvalue is lower for DCA than a CA ordination of the same data, since the CA eigenvalue includes

Table 7.5 Sample output from a DECORANA ordination. This was produced by PC-ORD, but much the same information is given by other implementations of this algorithm. Here the software output is listed in bold.

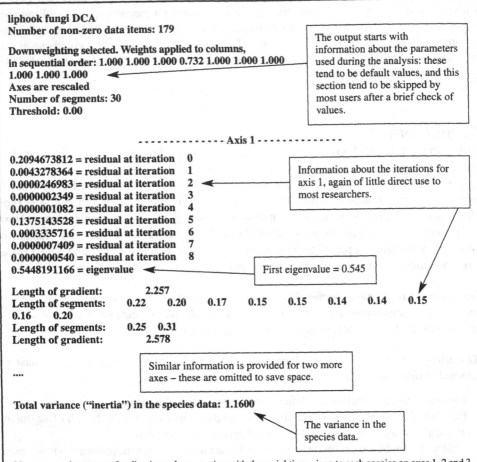

liphook fungi DCA
Number of non-zero data items: 179

Downweighting selected. Weights applied to columns,
in sequential order: 1.000 1.000 1.000 0.732 1.000 1.000 1.000
1.000 1.000 1.000
Axes are rescaled
Number of segments: 30
Threshold: 0.00

The output starts with information about the parameters used during the analysis: these tend to be default values, and this section tend to be skipped by most users after a brief check of values.

- - - - - - - - - - - - - Axis 1 - - - - - - - - - - - - -

0.2094673812 = residual at iteration   0
0.0043278364 = residual at iteration   1
0.0000246983 = residual at iteration   2
0.0000002349 = residual at iteration   3
0.0000001082 = residual at iteration   4
0.1375143528 = residual at iteration   5
0.0003335716 = residual at iteration   6
0.0000007409 = residual at iteration   7
0.0000000540 = residual at iteration   8
0.5448191166 = eigenvalue

Information about the iterations for axis 1, again of little direct use to most researchers.

First eigenvalue = 0.545

Length of gradient: 2.257
Length of segments: 0.22   0.20   0.17   0.15   0.15   0.14   0.14   0.15
0.16   0.20
Length of segments: 0.25   0.31
Length of gradient: 2.578

....

Similar information is provided for two more axes – these are omitted to save space.

Total variance ("inertia") in the species data: 1.1600

The variance in the species data.

Now comes the output of ordination values, starting with the weightings given to each species on axes 1, 2 and 3. Note that the middle column gives the eigenvalue for the first axis (0.545), then repeats the weightings, but this time gives them sorted into ascending order. The text below shows the first axis weightings, starting with *Paxillus involutus* (an early successional species) and ending with *Cortinarius semisanguineus* (a late successional species).

liphook fungi DCA

SPECIES SCORES				RANKED 1			RANKED 2		
N NAME	AX1	AX2	AX3	EIG=0.545			EIG=0.095		
1 BOLEFERR	120	-46	90	7 PAXILL_1	346		10 SUILLU_3	255	
2 CORTIN_2	-81	231	60	5 LACCAR_1	209		2 CORTIN_2	231	
3 GOMPHI_1	58	-118	130	4 INOCYB_1	158		9 SUILLUTE	195	
4 INOCYB_1	158	185	-83	1 BOLEFERR	120		4 INOCYB_1	185	
5 LACCAR_1	209	80	13	3 GOMPHI_1	58		6 LACTAR_1	170	
6 LACTAR_1	-74	170	153	8 SUILBOVI	47		5 LACCAR_1	80	
7 PAXILL_1	346	61	193	10 SUILLU_3	16		7 PAXILL_1	61	
8 SUILBOVI	47	27	79	9 SUILLUTE	14		8 SUILBOVI	27	
9 SUILLUTE	14	195	55	6 LACTAR_1	-74		1 BOLEFERR	-46	
10 SUILLU_3	16	255	-127	2 CORTIN_2	-81		3 GOMPHI_1	-118	

(Continued)

Table 7.5 (Continued).

liphook fungi DCA SAMPLE SCORES - WHICH ARE WEIGHTED MEAN SPECIES SCORES														
1	OBS	01	209	79	13	|	5	OBS	05	321	|	29	OBS	29	194	|
2	OBS	02	240	75	54	|	3	OBS	03	306	|	22	OBS	22	110	|

... (data omitted to save space)

| 34 | OBS | 34 | 83 | 52 | 57 | \| | 22 | OBS | 22 | 34 | \| | 24 | OBS | 24 | 9 | \| |
| 35 | OBS | 35 | 50 | 24 | 80 | \| | 29 | OBS | 29 | 0 | \| | 25 | OBS | 25 | 0 | \| |

****************************** Calculations finished *******************************

axis distortion which is removed by detrending. The author is unaware of any guidelines for assessing the importance of CA/DCA eigenvalues, although in principle a Monte Carlo approach could be used. Økland (1999) reviews the variance explained by CA and DCA axes, and criticises the use of eigenvalues to assess the effectiveness with which the ordination has recovered a meaningful axis in the data.

An alternative approach to assessing the importance of DCA axes (or, indeed, any other ordination) is to run an after-the-fact correlation between the interobservation distances in ordination space and the equivalent interobservation distances in the Euclidean space defined by the source data. This is offered by one major package (McCune and Mefford, 1999), but an applied example below highlights the problem that changes in the distance metric make dramatic differences to the correlation.

Returning to Table 7.5, it can be seen that the information about the species weightings is given twice. These are first given in the order that the species appeared in the source data matrix. Thus in the example of Table 7.5 the weightings for *Cortinarius semisanguineus* (abbreviated to Cs) are −81, 231 and 60 for axes 1, 2 and 3 respectively. As with PCA, the exact values of these weights are unimportant. What matters is the pattern of weights between species.

This same information is repeated for axis 1 in the right-hand side of the table, except that the species are now presented in order of decreasing weights.

In normal DECORANA output there would be two more columns, giving species in descending order for axes 2 and 3. The third column has been removed from Table 7.5 for clarity of presentation. Similarly, the original DECORANA output gave a second table of weightings in exactly the same format for the observations (here meaning the years of data collection). This has also been removed for the sake of simplicity.

7.3.3 DCA biplots

Just as with PCA, it is often informative to overlay two parts of a DCA ordination onto one graph known as a biplot. With a PCA the information presented consisted of sample scores and of eigenvector elements, and appreciating why this is valid requires some in-depth understanding of the mechanics of PCA. With DCA, biplots are more readily accessible, since the two sets of information that can be presented are simply the species scores and the sample scores. By convention sample scores tend to be plotted as coordinates while species tend to be plotted as arrows coming out of the origin. An example is given below.

7.4 Worked examples

7.4.1 The Liphook forest fungal dataset

The data from Table 1.9 were ordinated by DCA, and the results are shown in Table 7.2, with actual output in bold and explanatory comments in normal text.

The output starts with information about the operation of the algorithm: the values used and the progress of the iterations. This tends not to be examined closely by most researchers. It can be seen that the first eigenvalue was 0.545 while the overall inertia of the dataset was 1.16 (allowing a first approximation that the first axis accounts for 0.545/1.16 = 47.0% of the total variance). In fact there are currently no good tests for assessing the importance of DCA eigenvalues, but values over 0.3 are generally investigated further. An alternative tactic is to use an after-the-fact correlation between the distances in the ordination space and distances in the original data space. McCune and Mefford (1999) recommend this as a general tool to explore the effectiveness of any ordination, with the caveat that the choice of distance measure is crucially important, and that for DCA the measure recommended is relative Euclidean. In this case the correlation between ordination distances and distances in the original data is 0.802, suggesting that the ordination has effectively captured much of the structure of the data (Table 7.6), although it is not clear how many degrees of freedom this coefficient takes. Table 7.6 includes a warning over the sensitivity of this test to changes in the distance measure applied to the source data: altering

Table 7.6 Correlations between interobservation distances in DCA ordination space and the Euclidean space defined by the source data.

The correlation between distances in the ordinated and the source data, using the relative Euclidean distance function recommended by McCune and Mefford (1999)

Coefficients of determination for the correlations between ordination distances and distances in the original n-dimensional space:

Axis	R Squared Increment	Cumulative
1	.802	.802
2	.023	.825
3	.036	.861

Number of entities = 35
Number of entity pairs used in correlation = 595
Distance measure for ORIGINAL distance: Relative Euclidean

The same correlations for the same dataset, but using instead Euclidean distances in the source data matrix. Note that this apparently minor change has dramatically reduced the correlation coefficients.

Axis	R Squared Increment	Cumulative
1	.045	.045
2	.018	.064
3	−.009	.054

Number of entities = 35
Number of entity pairs used in correlation = 595
Distance measure for ORIGINAL distance: Euclidean (Pythagorean)

relative Euclidean to absolute Euclidean distance reduces the first axis correlation from 0.802 to 0.045.

The next step is to examine the species weightings. These are given twice in the same table, first (under **AX1 AX2 AX3**) in the order that the species appeared in the source data matrix and secondly (under **Ranked 1**) sorted into ascending order. The pattern on the first axis is for the early successional species *Laccaria laccata* and *Paxillus involutus* to have the highest weights, while the lowest weights are for late successional species such as *Cortinarius semisanguineus*.

The original DECORANA output gave a full second table of weightings for the observations. These values have been partially removed to avoid cluttering the table, but the layout used is identical to the species ordination.

As usual with ordination, obtaining the weightings is a start, not an end in itself. The next step is to inspect the ordination diagrams. These are shown in Figures 7.7–7.9 for the species only, sites only and biplot ordination diagrams respectively. A comparison between CA and DCA ordination of the same data is given by comparison of Figures 7.3 and 7.9. It can be seen that they exhibit the same basic pattern, but that CA has produced a rather 'black and white' division into early- and late-successional species, while DCA has separated species and sites more widely.

We can explore using an after-the-fact correlation between interobservation distances in ordination space and in the Euclidean space of the source data (McCune and Mefford, 1999). The correlations are listed in Table 7.6, showing that an impressive correlation of 0.802 is reduced to 0.045 by switching the distance measure applied to the source data from relative Euclidean to absolute Euclidean distance. This probably reflects the presence of a small number of multivariate outliers in the source data (notably plot 1 in 1990 – see figure 6.19), but shows that this approach to deciding on the importance of DCA axes is prone to instabilities.

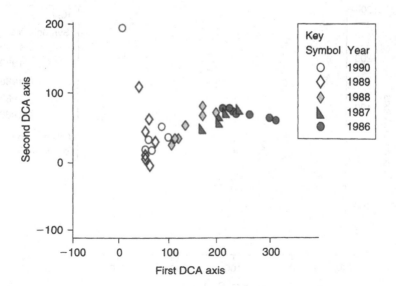

Figure 7.7 Detrended correspondence analysis of the Liphook pine forest dataset, showing only the site ordination (overlain with year).

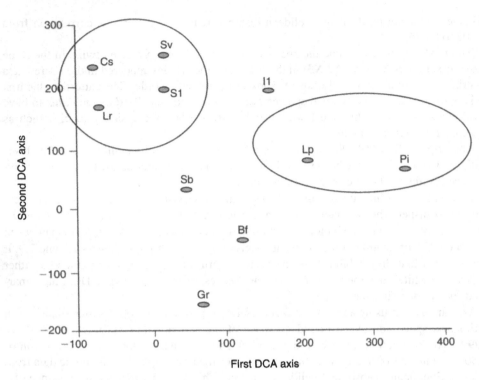

Figure 7.8 The DCA ordination of the Liphook pine forest fungal dataset, showing only the species ordination. Abbreviations are listed in Table 1.9. Ellipses identify early (Lp, Pi) and late (Cs, Sv, Lr, S1) successional species.

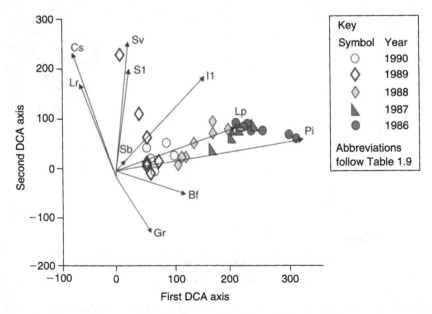

Figure 7.9 Detrended correspondence analysis of the Liphook pine forest dataset, showing both the site ordination (overlain with year) and the species ordination.

7.4.2 Microcrustacea in Alaskan streams

The data presented in Table 1.11 represent an ecological succession, so DCA would be a natural choice of ordination technique to analyse these. The log-transformed data in Table 1.15 were subjected to DCA (selecting the default options to downweight rare species), and the ordination is summarised in Table 7.7. These results are graphed in Figure 7.10, which

Table 7.7 The DCA ordination of Alaskan stream data (from Table 1.13).

	DCA axis		
	1	2	3
Eigenvalue	0.836	0.321	0.144
The Sites ordination			
stonefly	0.00000	0.00000	102.23492
stonefly	0.00000	0.00000	102.23492
wolf pt	601.66156	190.92270	125.19103
wolf pt	601.66156	190.92268	125.19103
wolf pt	601.66156	190.92268	125.19103
wolf pt	601.66156	190.92270	125.19103
wolf pt	601.66156	190.92270	125.19104
tyndall	385.99100	97.32552	91.12443
tyndall	428.39047	91.17555	87.53778
tyndall	467.70328	105.79726	103.54425
tyndall	428.40079	91.13747	87.58101
tyndall	426.47302	98.29483	79.45861
berg n	426.25150	99.11736	78.52521
berg n	375.10538	290.18307	155.30869
berg n	404.69800	198.03064	137.35307
berg n	407.61472	168.31296	0.00000
berg n	385.77310	251.28557	107.04556
berg s	457.10654	183.11516	151.59094
berg s	444.88831	29.92176	157.05042
berg s	480.30981	118.66619	101.05463
berg s	510.93579	97.75036	143.62827
berg s	476.82791	125.58781	91.98048
rush pt	410.18573	184.99582	218.62845
rush pt	601.66156	190.92270	125.19103
rush pt	407.61472	168.31294	0.00000
rush pt	494.24786	129.58188	101.38866
rush pt	407.61472	168.31296	0.00000
carolus	791.79883	214.43303	127.84300
carolus	703.72540	203.71646	127.07436
carolus	723.32745	206.60703	117.89250
carolus	746.14948	209.16505	117.80858
carolus	711.93665	205.19003	117.61579
The Species ordination			
lognito	773.95514	212.51979	128.37029
logatill	764.61847	212.48508	101.09552
logatid	375.48318	340.06989	280.20645
logmarin	0.00000	0.00000	102.23492
logbryhi	407.61472	168.31294	0.00000
logbryzs	444.88831	29.92176	157.05040
logacver	601.66156	190.92268	125.19103
logalong	833.47119	218.90125	126.61156
loggrapt	749.29535	210.73663	103.37088
logchydo	325.91461	369.62994	131.09727
logmacro	793.85950	215.30835	96.37854

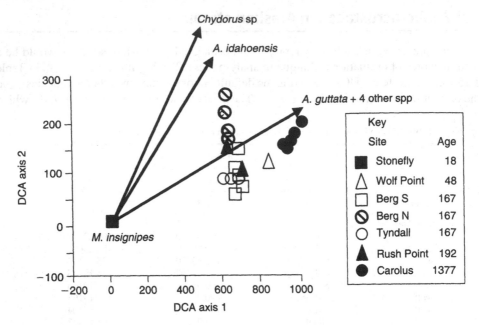

Figure 7.10 DCA ordination of the Alaskan streams dataset. The projection for *M. insignipes* lies exactly on top of the projection for the Stonefly site. Some observations and species have been removed to reduce clutter.

shows a pattern closer to the CA ordination of these data (Figure 7.3) than the PCA ordination (Figure 6.19). This is of course to be expected, since DCA is a modification of the CA algorithm. The depauperate community in the youngest stream (Stonefly) is shown on the first two axes of this ordination at (0,0), the extreme left-hand end of the first axis, as does the species *M. insignipes* (which was mainly found at this site). This zero projection ceases to be true on the third axis, where the Stonefly site has a score of 97.2. Data from the oldest stream lie at the opposite end of the first axis. In between come communities from the other sites in approximate order of increasing age, although observations from the 48 year old Wolf Point came far higher on the first axis than would be expected on the basis of age.

It is possible to use standard bivariate correlation to explore whether any of the DCA axes may be explicable in terms of environmental parameters, but there are pitfalls to beware of. The environmental data (given in Table 1.12) consists of one mean value per stream, while the DCA ordination took five observations per stream. It would be mechanically possible to compute the correlation between the five DCA scores for each site and five identical values for each physical measurement, but this would be utterly invalid – an example of pseudoreplication (Section 1.5.2). Instead it is first necessary to calculate a mean DCA axis score for each stream, and to correlate this with the physical parameters. There are several possible approaches to calculation of this single DCA score. Here we shall simply calculate the mean of the five DCA scores for each stream. An alternative would be to return to the source data, calculate a total of each species for each site, log-transform this total, and rerun the DCA ordination of these. (In fact the results given by this method are very similar.) Table 7.8 shows the correlation matrix between the DCA axes and the environmental

Table 7.8 Correlations (calculated by Spearman's correlation coefficient) between DCA axis scores (one mean per stream) and physical features of the Alaskan stream data. No correlations were significant at the 5% level.

	Age, years	Pfankuch index	%cwd	Temp, C	turbidity	conductivity	alkalinity
DCA1	0.57	0.64	0.04	0.25	−0.21	−0.18	−0.25
DCA2	0.51	0.00	0.64	0.29	0.46	0.51	0.39
DCA3	−0.40	−0.29	−0.04	−0.21	0.00	−0.16	−0.04

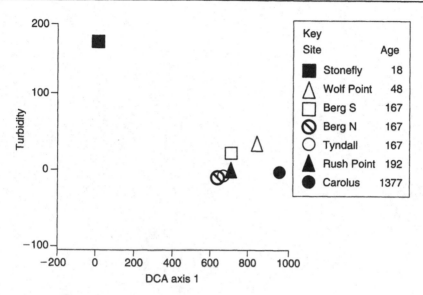

Figure 7.11 The relationship between the first DCA axis and water turbidity for the Alaskan stream data.

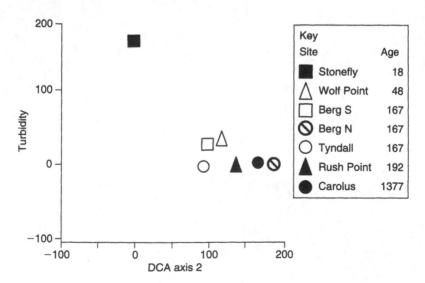

Figure 7.12 The relationship between the second DCA axis and water turbidity for the Alaskan stream data.

parameters, showing that the correlation between the first ordination axis and site age is weak, and that turbidity of the water appears to be a best correlate of both the first and second axes. Figures 7.11 and 7.12 show the relationship between turbidity and the first two DCA axes, showing the dominant trend to be that the youngest site (Stonefly) has a higher turbidity than the other sites.

8

Classification: cluster analysis and related techniques

'The availability of computer packages of classification techniques has led to the waste of more valuable scientific time than any other statistical innovation (with the possible exception of multiple regression techniques).' Cormack (1971).

'The best form of cluster analysis is ordination, because ordination is not a form of cluster analysis'. Byron Morgan, personal communication.

8.1 Introduction to classification

One of the recurring themes in multivariate analysis is the idea of seeking patterns within large blocks of data. As shown in the chapters on ordination techniques, one of the most informative patterns that can emerge is when data points appear to fall into discrete clusters. This happens when the individuals or observations comprising the cluster have similar attributes, suggesting that they should be classified together in some way (as a discrete sub-species, a habitat type, a socio-economic grouping, etc.).

Classification techniques are used specifically to search for such divisions within data, seeking to classify individuals (quadrats, samples, populations, etc.) on the basis of their attributes (species composition, chemical content, sociological characteristics, etc.). The aim of this chapter is to introduce some of the approaches used in the classification of multivariate data. The aim of classification techniques is to identify discrete subsets of individuals with similar characteristics. Such individuals can validly be thought of as forming clusters of points in some form of data space, and the whole field of multivariate classification is sometimes referred to as cluster analysis. This is not formally correct – cluster analysis is one particular set of approaches to this field, although by far the most commonly used.

Like ordination, classification techniques are intended to help a researcher explore data and generate hypotheses, and like ordination, there is an unfortunate tendency to rely on them as an end in themselves. It is important to understand that ordination techniques simply portray a representation of data, and may or may not show evidence of clustering. By contrast, classification techniques are specifically intended to divide data up and will always find clusters (even in random noise). Classification techniques are objective only in as much as they are repeatable: applying the same technique to the same dataset will usually give the

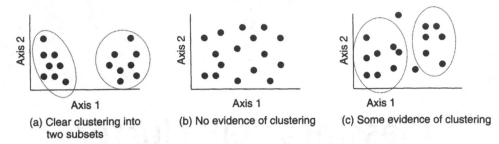

Figure 8.1 Examples of possible clustering patterns in ordination diagrams.

same result (although a few clustering algorithms – K-means clustering and an old technique called COMPCLUS – use random starting points for cluster formation and may give erratic results in poorly structured data). Unfortunately no classification technique supplies tests of significance, and different classification techniques may give very different classifications of the same data. Ultimately the effectiveness of any classification technique must be judged by a subjective assessment of its helpfulness to the researcher (Goodall, 1978).

One philosophical difficulty with the whole field of cluster analysis is that there is no formal, unambiguous definition of what is meant by a cluster. Consequently no formal rules can be laid down for identifying clusters, and ultimately there must come a subjective decision by the researcher about the clustering patterns within his/her data (Rao, 1952). Humans are good at locating clusters in visual patterns – rather too good in fact, and will normally identify clustering of points within a random scattering.

Figure 8.1 shows three hypothetical ordination diagrams, containing different levels of clustering. Figure 8.1(a) shows clear-cut clustering of the data into two different sets. In this case the distance between the clusters is larger than the greatest distance between any points within the cluster (a condition known as ball clusters). Figure 8.1(b) shows a continuum of points over the ordination diagram, with no indication of clustering and no reason to believe that the data contain discrete subpopulations. Figure 8.1(c) shows a more typical result for environmental data, with some evidence of clustering but also some data points lying between the two main groups.

Just as with ordination, many different numerical classification techniques were devised in the 1960s and 1970s as access to computing power became routine, but only a small number have survived the test of time to become standard research tools incorporated into modern statistical packages (Kent and Ballard, 1988). Currently most researchers seeking to classify a set of measurements would turn to cluster analysis or TWINSPAN without considering the extensive body of research that has been performed on devising useful classification algorithms. This chapter aims to introduce the widely used techniques, and also to mention a few of the less popular techniques which are of historical interest or conceptually useful.

8.2 A taxonomy of classification techniques

There are many approaches to numerical classification, of which the major divisions are summarised in Figure 8.2. The first division encountered is between **hierarchical** and **non-hierarchical** techniques. The vast majority of classification techniques in use

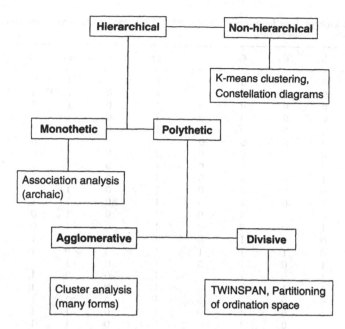

Figure 8.2 A dendrogram showing a taxonomy of classification techniques.

are hierarchical, which means that each division (cluster) can be identified as belonging to a higher-order cluster, allowing the construction of a **dendrogram**. This is a diagram which portrays patterns of clustering, and allows the relationships of each cluster to all the others to be seen at a glance. Figure 8.2 is itself a dendrogram. 'Dendrogram' literally means branching diagram, so called due to their similarity to the branching patterns of trees. It is however possible to create diagrams which show relationships between individuals without imposing such a hierarchy – the realm of non-hierarchical classification techniques.

8.2.1 Non-hierarchical classification techniques

Non-hierarchical classification techniques seek to subdivide data into categories (clusters) without attempting to explore how the clusters relate to each other, or to seek further structure within each cluster. As such they are rather crude tools, although potentially helpful for preliminary exploration of data. With the advent of powerful personal computers running clustering packages, this approach to classification has been almost wholly superseded by the spectrum of hierarchical techniques described below. Non-hierarchical techniques are worth mentioning for historical reasons, and because their outputs are conceptually simpler than dendrograms.

The simplest example of a non-hierarchical classification technique involves the construction of a **constellation diagram**. This is a figure which illustrates significant positive relationships between entities (usually species in an ecological community). Constellation diagrams have the virtues of being straightforward to produce without a computer, of only needing presence/absence data, and of being readily understood without recourse to

Table 8.1 The Liphook pine forest toadstool data 1986–1990, recoded as binary variables (0 = absent, 1 = present). Abbreviations as Table 1.11.

Year	Plot	Bf	Cs	Gr	Il	Lp	Lr	Pi	Sb	Sl	Sv
86	1	0	0	0	0	1	0	1	0	0	0
86	2	0	0	0	0	1	0	1	0	0	0
86	3	0	0	0	0	1	0	1	0	0	0
86	4	0	0	0	0	1	0	1	0	0	0
86	5	0	0	0	0	1	0	1	0	0	0
86	6	0	0	0	0	1	0	1	0	0	0
86	7	0	0	0	0	1	0	1	0	0	0
87	1	0	0	1	0	1	0	1	1	0	1
87	2	1	0	1	0	1	0	1	1	0	0
87	3	0	0	1	0	1	0	1	1	0	0
87	4	1	0	1	0	1	0	1	1	0	0
87	5	0	0	0	0	1	0	1	0	0	0
87	6	1	0	0	0	1	0	1	0	0	0
87	7	0	0	1	0	1	0	1	1	0	0
88	1	0	1	1	1	1	0	0	1	1	1
88	2	1	0	1	0	1	0	1	1	1	1
88	3	1	0	1	1	1	0	0	1	0	0
88	4	1	1	1	0	1	0	0	1	1	1
88	5	0	0	0	0	1	0	1	1	0	0
88	6	1	0	1	0	1	0	1	1	0	1
88	7	1	0	1	0	1	0	1	1	0	1
89	1	0	1	1	1	1	0	0	1	1	1
89	2	1	0	1	0	1	0	1	1	0	1
89	3	1	0	1	0	1	0	1	1	1	1
89	4	1	0	1	0	1	0	1	1	1	1
89	5	0	0	1	0	1	1	1	1	0	1
89	6	0	0	1	0	1	0	1	1	0	1
89	7	1	1	1	0	1	0	0	1	0	1
90	1	1	1	1	1	1	1	1	1	1	1
90	2	1	1	1	0	1	1	1	1	0	1
90	3	0	1	1	0	1	1	1	1	1	0
90	4	1	1	1	0	1	1	1	1	1	1
90	5	0	0	1	0	1	1	1	1	0	1
90	6	1	0	1	0	1	0	1	1	0	1
90	7	0	1	1	0	1	1	1	1	0	0

mathematical abstractions. These diagrams are not currently produced by any standard package, and are little used in mainstream research output. As such their main value is as a useful introduction to community description at an undergraduate level. It should be noted that this is a classification of species within samples, rather than the more usual approach of using species as attributes to classify samples.

The construction of a constellation diagram involves calculating the significance of the association between each pair of species in turn. This is usually done by a χ^2 statistic of association on presence–absence data (testing the null hypothesis that two species are randomly distributed with respect to each other), although a correlation coefficient could equally be used. Either way, only positive associations are used (those where the observed value for co-occurrence exceeds the expected value). An example used here is based on a presence/absence version of the Liphook forest dataset (listed in Table 8.1) and the analyses

Table 8.2 The construction of a constellation diagram.

The analysis here uses data from Table 8.1. For each pair of species in turn, a contingency table is set up giving the number of times that the species occur together or alone. The table for the two species *Gomphidius roseus* and *Suillus bovinus* is given as an example. This shows that of the 35 annual data sets, nine had neither species, one had *S. bovinus* alone, none had *G. roseus* alone, and 25 had both species.

Observed values:

	+Gr	−Gr	Sum
+Sb	25	1	26
−Sb	0	9	9
Sum	25	10	35

	+Gr	−Gr	Sum
+Sb	25 × 26/35 = 18.6	10 × 26/35 = 7.4	26
−Sb	25 × 9/35 = 6.4	10 × 9/35 = 2.6	9
Sum	25	10	35

This gives a χ^2 value of 30.3, and proves a significant association between the two species. This process is repeated for each pair of species, and the positive associations are assembled into a matrix. Only the lower half of the matrix is displayed below. * indicates a positive relationship significant at $p < 0.05$, ** shows a positive relationship significant at $p < 0.01$, and ! shows the leading diagonal. (An alternative approach to constructing this matrix would be to use a correlation coefficient to assess the extent of co-occurrence of species.)

	Bf	Cc	Cs	Gr	Il	Lp	Lr	Pi	Sb	Sl	Sv
Bf	!										
Cc	−	!									
Cs	−	−	!								
Gr	**	−	*	!							
Il	−	−	*	−	!						
Lp	−	−	−	−	−	!					
Lr	−	−	*	−	−	−	!				
Pi	−	−	−	−	−	−	−	!			
Sb	*	−	−	**	−	−	−	−	!		
Sl	−	−	*	*	*	−	−	−	*	!	
Sv	*	−	−	**	−	−	−	−	**	*	!

are given in Table 8.2. This shows how a matrix of association indices (here χ^2 significance values) can be assembled into a diagram which allows one to see which species are significantly co-associated (Figure 8.3). The constellation diagram shown in Figure 8.3 could be rescaled, inverted, or the relative positions of any of the species changed; its exact layout is arbitrary. All that matters is the pattern of linkages between species, which here shows little evidence of multiple communities. Kent and Coker (1992, p. 107) give an

Species names abbreviations are given in Table 1.9.

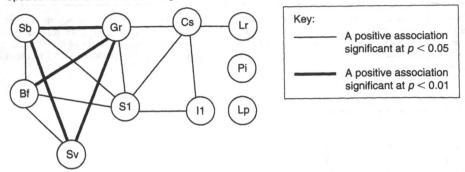

Figure 8.3 A constellation diagram summarising the association matrix listed in Table 8.2.

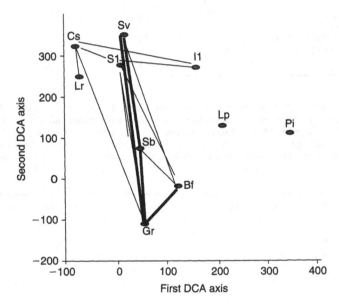

Figure 8.4 The DCA ordination of the Liphook pine forest fungal dataset overlain with a constellation diagram derived in Table 8.2.

example of a larger constellation diagram in which all species are interconnected but with some habitat preferences discernable. There are no rules to define how to lay out species within a constellation diagram, and in fact plotting an ordination diagram then overlaying the links is a good solution. An example of this is shown in Figure 8.4, where the DCA species ordination (originally given as Figure 7.8) is overlain with the constellation diagram of Figure 8.3.

A useful quick non-hierarchical clustering algorithm is **K-means clustering** (Sneath and Sokal, 1973; Anderberg, 1973). This involves the user specifying in advance the desired number of clusters to be identified. Either the cluster mid-points may be defined in advance by the user or suitable values for them may be found by iterative calculations by a computer.

The end result of a K-means cluster analysis will be the allocation to each individual in a dataset of a number identifying its cluster. Crucially, no information is given about how the clusters relate together because this is a non-hierarchical technique. K-means clustering may be useful for initial exploration of large datasets, since it is relatively undemanding of computer resources, but is rarely used in definitive published analyses.

For completeness it is worth mentioning a little-used non-hierarchical clustering technique known as **composite clustering** or COMPCLUS after the computer program which performed it (Gauch, 1979; 1980). This involves calculating a dissimilarity matrix (explained further in the following section), then randomly selecting a number of individuals to act as foci. Clusters are sought around these foci in two iterations, producing a small number of clusters. This technique is not guaranteed to reproduce the same clusters each time (due to the element of randomisation), and its main virtue was the speed of the algorithm. There have been a number of variants on this basic algorithm, including the computer programs CLUSLA (Louppen and van der Maarel, 1979) and FLEXCLUS (van Tongeren, 1986), and papers by Janssen (1975), Salton and Wong (1978) and Swain (1978). With increasing computer power this technique seems to have vanished entirely.

8.2.2 Hierarchical classification techniques

All the remaining classification techniques produce a dendrogram, which repeatedly subdivides a set of individuals into progressively smaller clusters until a stopping condition is encountered. The stopping conditions are defined by criteria such as each cluster containing just one observation, or a predefined number of subdivisions having been extracted.

The next subdivision of classification techniques (Figure 8.2) is between **monothetic** and **polythetic** approaches. The difference between these is that polythetic techniques use all the information within data, while monothetic techniques impose classification based on the presence or absence of one attribute at a time. Almost all modern techniques are polythetic, but it is worth being aware of the monothetic classification technique known as **association analysis**, since it was heavily used (especially in the study of plant communities) in the 1960s and early 1970s.

Association analysis dates from work by Goodall (1953) and was developed by Williams and Lambert (1959, 1960, 1961), and like constellation diagrams it relies on the construction of a matrix of χ^2 values testing association between variables (usually presence/absence of species). Instead of considering significance values, all the χ^2 values for one species are summed, and the species with the highest total is used to classify the data. The data are divided into two sets; those observation with the indicator species, and those without. The process may be repeated for each of the new subgroups until some limit is met. The usual stopping rule is that analysis of a group terminates when the largest single χ^2 value it generates is less than the square root of the number of observations within the group. Association analysis is hardly used now, having been superseded by superior techniques described below. It suffered particularly from high levels of misclassification, since each division is defined absolutely by the presence/absence of one species. The inevitable noise present in environmental data means that this is really not a satisfactory approach to classification (Hill, Bunce and Shaw, 1975; Orloci, 1978).

The remaining classification techniques to be considered use all the variables simultaneously in splitting up data into categories – they are said to be polythetic. These include the two most commonly used classification techniques: cluster analysis and TWINSPAN. Following the taxonomy of Figure 8.2 we will first consider the procedures which build up clusters from individual observations, the family of agglomerative methods collectively called cluster analysis.

8.3 Cluster analysis

8.3.1 Overview

Cluster analysis (sometimes called similarity analysis) is so widely used (and misused) as to be the default classification technique for some researchers. Various clustering algorithms are implemented on many standard statistical packages (including SAS, GENSTAT, SPSS, PC-ORD, MVSP, and many others). In addition there exists a package called CLUSTAN (originally a mainframe program, now available for PCs) whose main function is to undertake many different types of cluster analysis (Wishart, 1987). The inexperienced user may be forgiven for being perplexed by the variety of choice, and may expend a great deal of time exploring all available permutations of the available techniques. Unlike the techniques explained under Section 8.2.1, cluster analysis is both a mainstay of the research community and allows (indeed requires) a great many different choices to be made by users. Consequently, the aim of this section is to not merely to give an introduction to this family of techniques but to highlight some generic weaknesses and dangers in this form of data analysis.

8.3.2 The basic procedure

It is important to realise that there are many different procedures (algorithms) that can be used to produce a valid dendrogram. They are all considered to be the same method – cluster analysis – although different algorithms can find different patterns within the same data, and any one cluster diagram can validly be presented in many ways. Consequently cluster analysis suffers from the problem of a proliferation of possible results from any set of data. A similar problem occurs with Bray–Curtis ordination, but with cluster analysis the number of allowable permutations is even larger, and the interpretation of the results becomes correspondingly more difficult.

The general procedure is shown in Figure 8.5. It has many similarities to Bray–Curtis ordination (Chapter 5), although cluster analysis is widely used while the latter is almost extinct as a research tool. The starting points for any cluster analysis is a rectangular matrix describing measured properties of a set of individuals. Generally the data should be continuous, but this depends on the distance measure chosen, and even nominal data can be used providing that the distance measure used is based on percentage similarity, such as the Sørensen index (equation 5.3).

The next step is to create a square matrix of dissimilarities (or distances) between the individuals. Exactly the same procedure was followed with Bray–Curtis ordination (Chapter 5). Casual inspection of the matrix of dissimilarities allows one to identify pairs of individuals which have closely similar composition, since these will have low dissimilarity scores. The idea of cluster analysis is to build up a dendrogram of relationships between all the individuals, based on their similarities (as measured in the matrix of similarities). It is

1. Create a matrix of all the distances between each pair of observations.

2. Find the lowest distance in the matrix (excluding the zeros on the leading diagonal), identify which pair of observations generated this, and fuse these two observations together to form the first cluster.

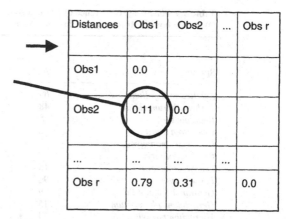

Distances	Obs1	Obs2	...	Obs r
Obs1	0.0			
Obs2	0.11	0.0		
...	...	...	...	...
Obs r	0.79	0.31		0.0

3. Develop the dendrogram (of which only a small part is shown here) by simple rules which allow three operations: Adding individuals to an existing cluster, create a new cluster, or fuse existing clusters, until all individuals are connected. The precise form of the rules defines the clustering algorithm used.

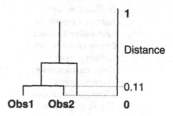

Figure 8.5 Steps in the construction of a cluster diagram.

important to understand that there is never one single correct dendrogram, but a large (potentially huge) number, depending on choices made about the rules for assembling the dendrogram. An analogy may be drawn with the calculation of diversity indices (Chapter 2). Although Hill (1973a) gave a function generating infinitely many different indices, in practice only a small number are actually used. Similarly for cluster analysis, Lance and Williams (1967) give a general function for generating hierarchies of clusters whose four parameters define infinitely many different clustering algorithms, but in practice only a small number of clustering algorithms are actually used.

8.3.3 A worked example

It would be useful to take a simple example to illustrate the clustering procedure. The data in Table 8.3 consist of the percentage frequency of 22 plant species in the communities from four areas in a disused quarry in North Yorkshire (Usher, 1975). Our aim is to use the floral data to identify patterns of similarity between these four areas.

The first step in a cluster analysis is to calculate some index of dissimilarity between the four areas. The index to be used here is known as the Euclidean distance, which relates to the positions of the four areas in data space. If the data for just two species – *Achillea millefolium* and *Arrhenatherium elatius* – are considered, they define a two-dimensional data space. The four areas can be plotted in this data space, as shown in Figure 8.6. It is now possible to calculate the distance between each of the points using Pythagoras' theorem; thus the distance between areas A and B in Figure 8.6 is found by:

$$((69 - 8)^2 + (48 - 21)^2)^{1/2} = (4450)^{1/2} = 66.71$$

Table 8.3 Percentage occurrence of 22 plant species in four areas (A, B, C and D) of the floor of a disused quarry in North Yorkshire. Data from Usher (1975).

Species	Area			
	A	B	C	D
Achillea millefolium	69	8	36	13
Arrhenatherum elatius	48	21	96	96
Briza media	5	6	12	0
Centaurea nigra	2	4	26	41
Crataegus monogyna	0	0	15	0
Crepis capillaris	0	10	0	0
Dactylis glomerata	15	2	16	22
Festuca ovina	96	95	98	15
Gentianella amarella	9	8	0	0
Heracleum sphondylium	15	13	6	1
Hieracium pilosella	74	79	0	0
Holcus lanatus	15	3	6	7
Lathyrus pratensis	0	0	0	17
Leontodon hispidus	41	75	8	0
Linum catharticum	13	2	0	0
Lolium perenne	0	0	0	15
Lotus corniculatus	83	81	2	7
Ononis repens	11	0	98	71
Plantago lanceolata	21	4	1	0
Tussilago farfara	0	10	0	0

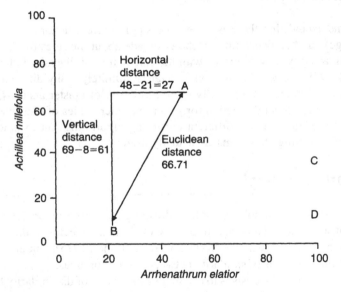

Figure 8.6 The construction of a two-species data space using the data of Table 8.3 to illustrate the calculation of a Euclidean distance.

The same approach can be used to find the distance between each pair of sites in the 22-dimensional data space defined by the data in Table 8.3.

As usual with multivariate techniques these calculations are tedious and usually undertaken by computer, but in order to understand the algorithm it is important to know what procedure the computer is following.

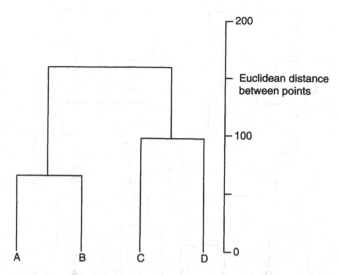

Figure 8.7 The construction of a dendrogram using the distance data of Table 8.7.

The Euclidean distances between the four communities A, B, C and D using all 22 variables are as follows:

	A	B	C	D
A	0.0	81.8	160.5	173.4
B		0.0	184.6	192.9
C			0.0	97.0
D				0.0

As with the dissimilarity matrices discussed in Chapter 5, and correlation matrices of Chapter 6, this matrix is symmetrical about the leading diagonal so is said to be triangular. The lower half of the matrix has been omitted for clarity.

These distances are an index of how dissimilar the communities are, and show that areas A and B are most similar while areas B and D are most distant. The clustering can now be started by joining up all the points which are separated by a Euclidean distance of less than 90 units (Figure 8.7). This rule only joins together areas A and B. If the threshold for joining points is raised to 100 units, areas C and D join up but remain distinct from the cluster AB. Finally, by raising the threshold to 161 units, A and C join up thereby linking all four sites.

This sequence can be built up as a dendrogram, as shown in Figure 8.7. The height of the dendrogram is equal to the greatest distance in the data (in this case 192.9, between areas B and D). Areas A and B join up at 81.8 units, forming the first cluster (shown as a horizontal line). The communities C and D join up at 97 units, forming a second cluster, then these two clusters merge at 160.5 units, where A and C join up. Notice that this leaves a 'tail' (root might be more appropriate) of some 30 units up to the maximum distance of 192.9 units.

An important point to remember about all dendrograms is that they are like children's mobiles, and may validly be rotated about any of the links. There are eight different ways of presenting the simple dendrogram in Figure 8.7, shown in Figure 8.8. For more complex dendrograms the number of permutations increases exponentially. Formally the number of

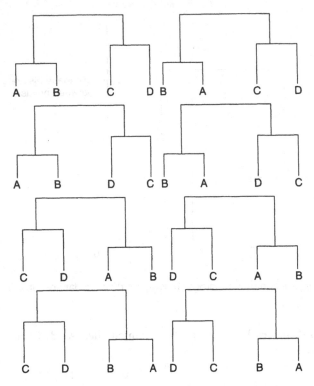

Figure 8.8 The eight different ways of presenting the dendrogram presented in Figure 8.7.

dendrograms which can be plotted for a dataset with N points is 2^{n-1}. In practice this means that interpretation of a dendrogram cannot rely on the sequence of observations along the bottom of the dendrogram.

The cluster analysis described above involved two decisions that need examining. The first was the distance measure, the second was the rule by which clusters were formed.

8.3.4 The choice of distance measure

The measure used was the Euclidean distance between points in data space. This is a natural distance measure, and the most commonly used in cluster analysis. It has the limitation that it cannot be used on nominal data, and (unless the data have been standardised in some way, such as normalisation) gives distances between data points that vary according to the units of measurement used. Several alternatives are available, of which the commonest are listed in Table 8.4. The CLUSTAN package offers over a dozen different distance measures. Some of the measures (the Jaccard, Sørensen and Czekanowski indices) use frequency of co-occurrence of species and can only be used with presence/absence or nominal data. There can be problems with using these indices in combination with certain clustering rules (described below) that involve iteratively recalculating intergroup distances as Euclidean (Wishart, 1969).

It is sometimes possible to create a unique distance measure suitable for a particular circumstance, subject to the constraint that the distance measure cannot go negative. Morgan (1981) gives an example from psycholinguistic research where subjects were played

Table 8.4 Some measures of dissimilarity frequently used in cluster analysis. This list is not comprehensive, but encompasses the majority of approaches used in analysing environmental data. In principle, distance measures listed here and those listed in Table 5.1 can be used interchangeably.

Euclidean distance. The most commonly used measure, which gives the actual distance between data points in data space. This is calculated as:

$$\text{Distance}_{AB} = \sqrt{\sum_{i=1}^{i=N} (X_{iA} - X_{iB})^2} \qquad (8.1)$$

where
Distance_{AB} is the distance between sites A and B in N-dimensional data space,
X_{iA} is the value for variable i at site A,
X_{iB} is the value for variable i at site B.

City-block distance. This is related to Euclidean distance, but instead of drawing a line directly between the two coordinates in data space the line follows a series of 90 degree bends, with each segment lying parallel to one of the axes. The result is like navigating around a series of blocks in a rectangularly organised city (such as New York). Mathematically the distance between individuals A and B is given by

$$\text{City-block distance}_{AB} = \sum_{i=1}^{i=N} |(X_{iA} - X_{iB})|$$

where
$|(x)|$ means the absolute value of x (ignoring its sign),
X_{iA} is the value for variable i for individual A,
X_{iB} is the value for variable i for individual B.

This is also known as the Manhattan distance.

Euclidean and city-block distances are special cases from an infinitely large family of distance measures derived from the Minkowski k metric:

$$\text{Distance}_{AB} = \left(\sum_{i=1}^{i=N} \left(|(X_{iA} - X_{iB})^k| \right) \right)^{1/k}$$

Substituting $k = 1$ gives the city-block distance; $k = 2$ gives the Euclidean distance. In practice higher values of k are little used.

Jaccard, Sørensen and **Czekanowski** dissimilarity coefficients. These were all introduced in Table 5.1. These have the useful feature of only requiring that a trait be present/absent in a pair of individuals, so allow cluster analysis of nominal data where distance measures cannot be calculated. It is worth noting that Wishart (1987) offers 12 distance measures, while Podani (1990) lists 31!

---

recordings of letters of the alphabet while listening to white noise. The letter perceived by the listener was recorded, to study auditory confusion between letters. These results were summarised as a square matrix of percentage mishearings of letter pairs, which acted as a distance matrix that could be directly entered into a cluster analysis.

8.3.5 Rules for cluster formation

The sequence of cluster formation shown in Figure 8.7 is an example of the simplest type of cluster analysis. It is known as **single-link clustering** or **nearest-neighbour clustering**, and is defined by clusters fusing on the basis of the smallest distance between any pair of their component individuals. This has the problem that two individuals may end up in the same cluster despite being widely separated, if they are linked by a chain of closely connected points. This effect is known as **chaining**. Another feature of a single-link dendrogram is that its constituent clusters tend to increase in size gradually, with each fusion adding just

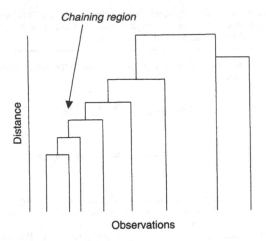

Figure 8.9 An example of a highly chained dendrogram, caused by an algorithm which tends to add observations to existing clusters one at a time.

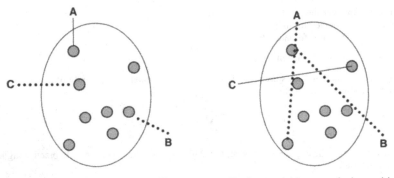

Nearest neighbour analysis would join up A next, since it is closer to its closest cluster member.

Farthest neighbour analysis would join up C next, since it is closer to its farthest cluster member.

Figure 8.10 An illustration of contrasting rules for cluster formation. The grey dots represent an ordination diagram of a set of individuals, some of which have previously been identified as being in a cluster (circled). Observations A, B and C are now considered for inclusion in the cluster – which will be included next?.

one or a small number of new elements. This produces inconclusive clusters that may be hard to interpret (a particular problem for large data sets). A highly chained single-link dendrogram is shown in Figure 8.9. It is sometimes possible to identify a dendrogram as being single link simply by the high degree of chaining present. The tendency for single-link cluster analysis to produce chained dendrograms has a positive feature, in that random data will not produce clear clusters. Other clustering techniques with higher resolution are capable of producing apparently clear clusters from random noise.

There are other ways of defining when clusters may be merged. **Complete-link clustering** allows fusions on the basis of the greatest distance between members of clusters, in contrast to single-link clustering which uses the lowest distance. This is illustrated in Figure 8.10. Complete-link clustering has the opposite problem to single link in that two individuals which are similar may end up in widely separated clusters, being held apart by a third individual which is close to one but distant from the other. A commonly used compromise is **minimum variance clustering** or **Ward's technique**, which generally gives results intermediate between single-link and complete-link clustering. (The 'default' clustering algorithm used by ecologists at present appears to be to apply Ward's method to a matrix of Euclidean distances, although users should be encouraged to explore several options to see which is most revealing.)

These and other approaches to defining clusters are summarised in Table 8.5. It is not necessary (or desirable) for the inexperienced user to learn the details of the plethora of clustering techniques available, although it is worth being aware of the incompatibility between some of the clustering techniques and some of the distance measures. The illegal combination most likely to be used by the unwitting is the application of minimum variance clustering to the Sørensen dissimilarity index.

8.3.6 Interpretation of dendrograms

There are no objective rules for interpreting dendrograms! The prudent researcher will repeatedly analyse the same dataset using different distance measures and clustering algorithms, looking for consistent patterns of co-occurrence. The intention should be to use the dendrograms to assist with hypothesis formation, and to this end the user will seek divisions that coincide with existing knowledge about the data. It is often useful at this stage to consider the background information about the individuals and the data collection techniques, the metadata (Section 1.5.2). The realisation that one cluster of points coincides with the data collected on one day, or by one particular research assistant, should ring alarm bells!

Naturally the most important divisions are those at the highest levels within the dendrogram, which should identify the same broad clusters under a range of clustering algorithms. There will always be a number of points whose position is uncertain and change affiliation according to the method used, hence the danger of basing groups simply on cluster analysis. These points may represent interesting transitional stages (known as **ecotones** in phytosociology), or may sometimes be discarded to artificially strengthen group structure. Kent and Coker (1992) suggest that when tentative group classifications have been derived they should be overlaid on an ordination diagram of the same data – known as **complementary analysis** (Kent and Ballard, 1988). An example of this overlaying is shown in Figure 8.11.

It is worth being aware that the number of possible cluster analysis algorithms offered on a statistical package may be impractically large, and the user will probably not want to explore all combinations. This arises because users can choose from several different dissimilarity measures and from many different clustering rules, and to consider whether data transformations are needed prior to undertaking the analysis. There are two common choices for dissimilarity (the Sørenson index and Euclidean distance) and four commonly used clustering rules (single link, minimum variance, UPGMA and complete link), making eight basic clustering algorithms. If Euclidean distance is chosen, the user may need to consider data transformations to prevent variables with the highest variance from dominating the analysis. Considerations of data transformations (typically square root or logarithmic) increase the number of combinations of clustering algorithms further. Adding some of the less frequently used distance measures and clustering rules gives over 50 clustering

Table 8.5 Approaches to the formation of clusters. This list gives all the commonly used techniques, although Lance and Williams (1967) give a general function that covers most of this list and allows the derivation of infinitely many other possible clustering algorithms. Note that some techniques should not be used on dissimilarity indices using percentage co-occurrence (such as the Jaccard or Sørensen indices).

Single-link (nearest-neighbour) clustering (Sneath and Sokal, 1973). This allows fusion of clusters based on the lowest value of the minimum distance between cluster members (Figure 8.10). It suffers from chaining, whereby individuals that are actually distant may end up in the same cluster if linked by a chain of closely connected individuals. Highly chained dendrograms tend to be hard to interpret, especially for large datasets.

Complete-link clustering (Sneath and Sokal, 1973). This allows cluster fusion based on the lowest value of the greatest distance between the members of two clusters (Figure 8.10). It suffers from two problems. Firstly, two closely similar individuals may end up in different clusters if they are both close to other individuals that are themselves highly dissimilar (in other words the clusters derived may not be well separated). Secondly, if there are ties in the dissimilarity matrix the dendrogram may not be unique.

Average linkage or **group average** clustering (Sokal and Michener, 1958). This is similar to single- and complete-link clustering, but allows cluster fusion based on the lowest value of the average distance between clusters. The average may be weighted in various ways – the commonest technique uses the simple unweighted mean of the distances, and is known as the **unweighted pair-groups method using arithmetic averages** or **UPGMA** clustering. Introduction of a weighting gives the **median** or **weighted pair-groups method using arithmetic averages (WPGMA)** clustering (Grieg-Smith, 1983). This latter technique should not be used on dissimilarity indices such as the Sørensen.

Minimum variance clustering, also known as **Ward's method** (Ward, 1963). This allows fusion of clusters on the basis of minimising within-cluster variance. It lies intermediate between single-link and complete-link clustering, and is probably the most widely used single technique, although (like complete-link clustering) ties in the dissimilarity matrix may mean that the dendrogram is not unique. It should not be used on the Sørensen or related indices.

Centroid sorting (Sokal and Michener, 1958) is similar to average linkage clustering, but after each fusion of two individuals into a cluster, these two individuals are removed from the raw data matrix and replaced by a new individual whose attributes are the average of its two members. The entire dissimilarity matrix (with one fewer members than before) is then recalculated. The process is repeated until all the dataset consists of one composite individual. This method is little used now due to the odd effect that after fusing A and B into a new individual (AB), the fusion between (AB) and C can sometimes take place at a *lower* level of dissimilarity than the fusion between A and B. These occurrences are known as reversals, and are contrary to the whole idea of dendrograms (Clifford and Stevenson, 1975). Cluster analyses that contain reversals may still be presented as dendrograms by replacing dissimilarity by a measure of information content known as Wishart's objective function (Wishart, 1969).

Flexible beta clustering (Sneath and Sokal, 1973). The latter involves the user entering a parameter between −1.0 and +1.0. A value of −1.0 gives a high level of chaining, approximating to single-link clustering, while +1.0 give a low level of chaining (approximating to complete-link clustering) and intermediate values give intermediate degrees of chaining. A value of 0.25 gives results similar to Ward's method. This should not be used on Sørensen or related indices.

Other occasionally used methods include **McQuitty's hierarchical clustering** and **Information analysis** (Williams *et al.*, 1966; Kent and Coker 1992). This technique was derived specifically for vegetation analysis and was widely used by Australian phytosociologists in the 1960s and 1970s but now has been superseded by other clustering techniques and by TWINSPAN. It uses an index of the information content of a cluster, with high information contents equating to large differences between individuals. Cluster fusions are allowed on the basis of the minimum increase in information content.

algorithms, while some of the options (using a generalised version of Minkowski's K metric for dissimilarity, or the flexible beta method for the clustering rule) require the user to enter a number which can take any of infinitely many values! In theory any change in the clustering algorithm may change the resulting dendrogram, and there are no significance

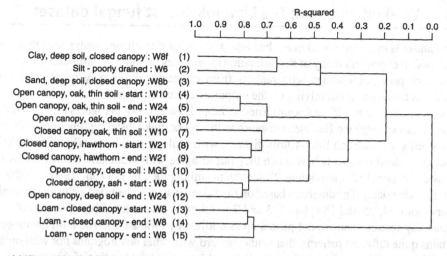

(a) Hierarchical dendrogram produced from the flexible clustering method. A description of each cluster is given followed by NVC community.

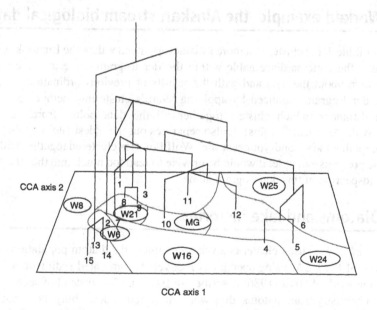

(b) The dendrogram of (a) showing the orientation of the clusters when superimposed onto the CCA ordination of the first two axes. The ends of the dendrogram are located in the most central point of the samples for each cluster.

Figure 8.11 An example of overlaying a dendrogram on an ordination diagram. From Vickers (2001, reproduced by permission of the author).

tests available to guide the user as to which dendrogram is the most useful or reliable. Consequently cluster analysis should be seen as a tool for data exploration and hypothesis generation, but never as an end in itself. Specifically the reader is cautioned against spending many days examining the output of all the clustering algorithms on offer in their software!

8.3.7 Worked example: The Liphook forest fungal dataset

This dataset is included not so much because it lends itself to cluster analysis, as that it has been used to exemplify many of the ordination techniques previously introduced, so a comparison of previous findings with outputs from cluster analyses may prove instructive (if only as a caution about relying on the output of cluster analyses).

Dendrograms for the 35 data points listed in Table 1.9 are shown in Figures 8.12–8.14, containing (respectively) the Euclidean distance with nearest-link clustering, Euclidean distance with Ward's method, and Bray–Curtis distance with farthest neighbour. What is remarkable about these dendrograms is how much they fail to agree with ordination techniques (which consistently identified observations 22 and 29, meaning plot 1 in 1989 and 1990, as different from the other data). The diagrams based on Euclidean distance identified a cluster containing observations 24, 25 and 28 (plots 2, 3 and 7 in 1989) – data points which have no obvious outstanding features and are not picked up as outliers by ordinations. The third dendrogram contains quite different patterns, that neither accord with other dendrograms nor with ordination techniques. The reader is referred to Cormack's quote at the heading of this chapter!

8.3.8 Worked example: the Alaskan stream biological dataset

These data (Table 1.11) cluster in a more satisfactory manner than the Liphook fungal data, meaning that the patterns discernable within the dendrogram agree to some extent with what is known about the data and with the results of previous ordinations. Figure 8.15 shows the dendrogram produced by applying Ward's clustering method to a matrix of Euclidean distances, which clusters together all the data points from three streams (Stonefly, Wolf Point and Carolus). It also separates out the oldest site (Carolus) from the others, except that the second-youngest site (Wolf Point) is clustered together with Carolus. As with the previous example, it would be unwise to read too much into the structure of the data from inspection of this dendrogram alone.

8.3.9 Diatoms and lake histories

Stoermer *et al.* (1993) used cluster analysis to examine the diatom populations within the Great Lakes of the USA, using specimens recovered from dated sediment cores. This is similar to the work of Dixit (1986) described in Example 3.4. Instead of seeking to reconstruct lake chemistry from diatoms, they were interested in describing the variation within the communities over the last 100 years. The cluster analysis used an average clustering algorithm (presumably UPGMA, although this is not explicitly stated) on Euclidean distances. The dendrogram is shown in Figure 8.16, and shows the highest level divisions to separate off four outlier samples; O5 (recent samples from Lake Ontario), M2 (Lake Michigan in 1877), H2 (Lake Huron, 1910) and E4 (recent samples from Lake Erie). These aberrant communities were hypothesised to be the result of man-made disturbance affecting water quality, with eutrophication in the recent samples and fires/land clearance explaining the older samples. Apart from these disturbed communities, the dendrogram generally clusters samples from the same lake together (most clearly for Lake Erie).

This example is included because it contains a warning about the interpretation of dendrograms. The text suggests that the dendrogram shows a gradient of eutrophication

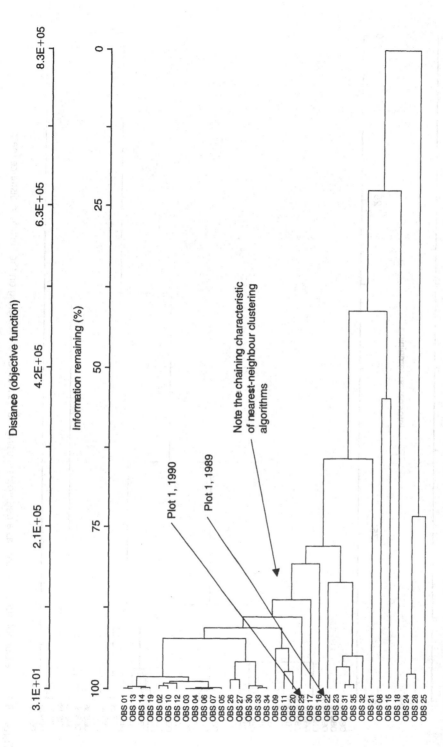

Figure 8.12 The Liphook fungal dataset as a dendrogram using nearest-neighbour clustering on the Euclidean distance matrix. Note that data from plot 1 in 1989 and 1990, which were identified as outliers by PCA and CA, are now buried in central positions.

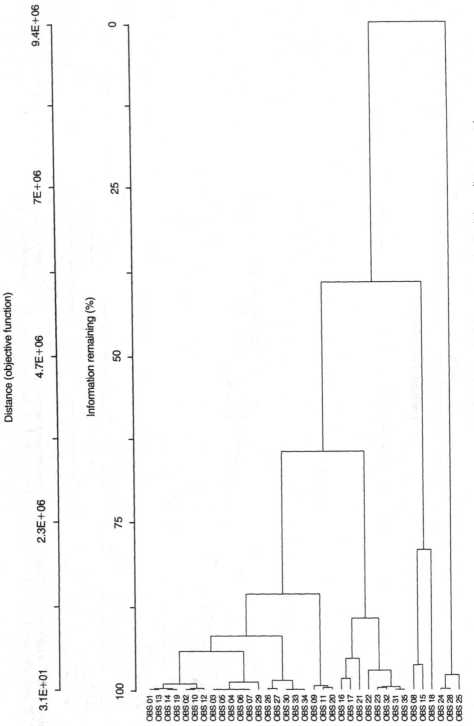

Figure 8.13 The Liphook fungal dataset as a dendrogram using Ward's method of clustering on the Euclidean distance matrix.

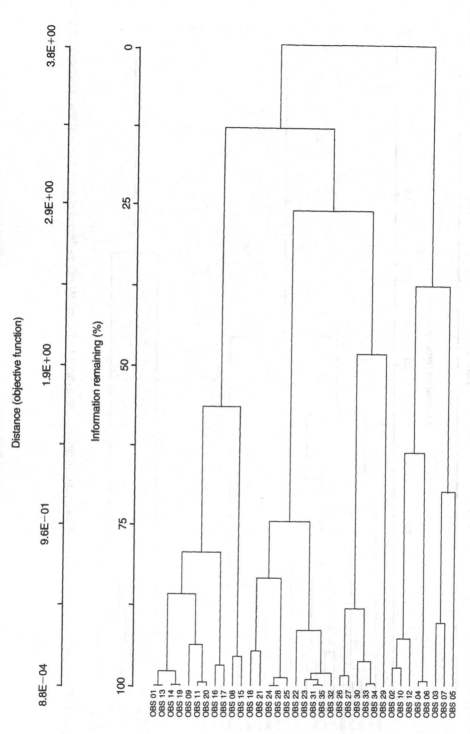

Figure 8.14 The Liphook fungal dataset as a dendrogram using farthest-neighbour clustering on a matrix of Bray–Curtis (Sørensen) distances.

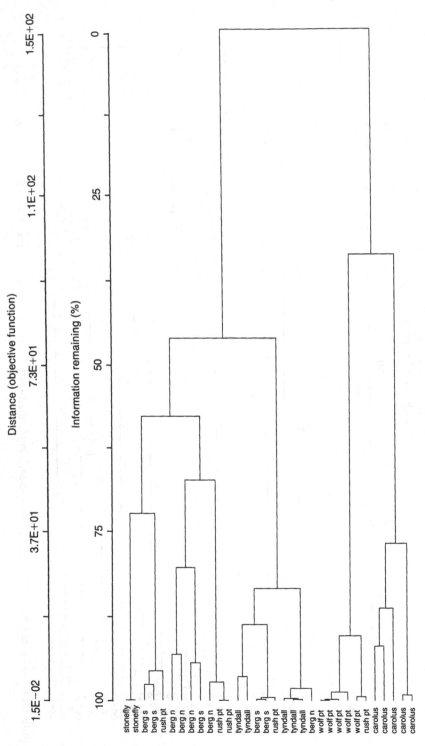

Figure 8.15 The dendrogram showing the Alaskan stream biological data (Table 1.11) using Ward's method of clustering on a matrix of Euclidean distances.

with samples from the eutrophic Lake Superior at the top of the figure and the oligotrophic Lake Erie at the bottom. Unfortunately dendrograms are not like ordination spaces, and do not contain gradients. They are like children's mobiles, able to be rotated about any link (Figure 8.8 shows a simple example). The dendrogram of Figure 8.16 could be redrawn equally validly with the Lake Superior samples split up, displaying a very different pattern. There are over 67 million different ways of drawing this dendrogram, all equally valid!

8.4 Divisive classification techniques

Here we examine the approaches to classification which take an entire dataset and progressively divide it into categories.

8.4.1 Subjective division of ordination space

This approach to classification comes directly out of the interpretation of ordination diagrams (Chapter 4). When an ordination technique produces evidence of clustering, it is legitimate to impose subjective boundaries on the diagram as an aid in classifying data (Roux and Roux, 1967; Prentice, 1980). This is a hierarchical technique since each subjectively defined cluster can be further split, if the data appear to merit subdivision. Figure 7.2 shows an example of this, since it visually isolates two groups of species as clearly outside the pattern described by the remainder of the data.

There are several possible ways to decide on the location of boundaries, of which the simplest is to run them along regions with low density of data points. Wildi (1979, 1980) gives algorithms for locating boundaries in ordination space, but their apparent objectivity is illusory since the choice of which procedure to use is still subjective. Division of ordination space usually leaves a few data points with an uncertain status, having been allocated to a cluster by a process approximating to guesswork. The less clear the data, the less satisfactory is this subjective approach. On the positive side, it forces the researcher to admit the existence of an element of uncertainty in cluster boundaries. This uncertainty should always be present, but can be deceptively lacking in the output of a standard clustering package.

8.4.2 Two-way indicator species analysis (TWINSPAN)

Background
This widely used technique comes from the same stable as DECORANA, being a FORTRAN program written by Mark Hill of the Institute of Terrestrial Ecology, to facilitate vegetation analysis (Hill, 1979b; 1994). Just as DECORANA has become an accepted acronym for detrended correspondence analysis, TWINSPAN has become a widely used acronym for two-way indicator species analysis. This technique is an extension of indicator species analysis (Hill, Bunce and Shaw, 1975), but is a great improvement over the original method which was monothetic and therefore prone to misclassification. TWINSPAN takes a dataset and progressively divides it up using all the information content of the data, so is said to be a polythetic divisive classification technique. This contrasts with cluster analysis, which progressively merges individuals so is said to be a polythetic agglomerative technique. Thus with TWINSPAN we complete the taxonomy of classification techniques

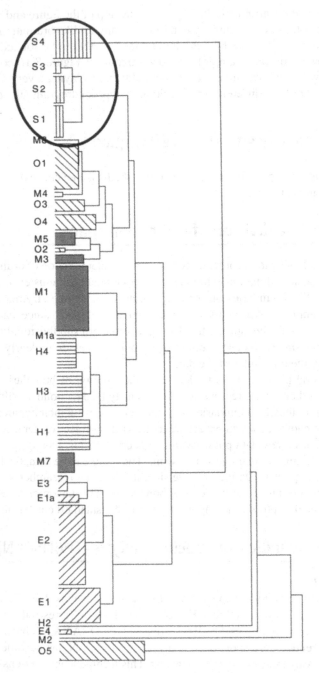

Figure 8.16 A dendrogram analysing diatom communities from the great lakes. Although the top four sites all come from the same lake (Superior), the algorithm has in fact placed S4 a long way away from S1-S3, and their being adjacent in this diagram is coincidential. Cluster dendrogram of samples in cores from the Laurentian Great Lakes. Prefix and pattern of bars indicate lake from which samples were derived: E – Erie; H – Huron; M – Michigan; O – Ontario; S – Superior. Reproduced by kind permission of the American Society of Limology and Oceanography.

(Figure 8.2). It is worth adding that TWINSPAN is probably the most reliable general classification technique of all those listed in Figure 8.2.

TWINSPAN classifies individuals and their attributes together into a two-way table. This contrasts with all the classification techniques mentioned so far, which either classify individuals on the basis of their attributes or (performing an inverse analysis) classify attributes on the basis of individuals – but without there being any relationship between the two classifications.

TWINSPAN tables

One of the most useful features of TWINSPAN is the two-way ordered tables that it produces. These tables order samples along a gradient (the first axis of a correspondence analysis ordination – Chapter 7), and simultaneously order attributes (usually species) along an equivalent gradient, showing the frequency of each species in each sample, and in addition supplying a dendrogram that dichotomously classifies samples and another that classifies species. This is a wealth of information, and this explains why TWINSPAN has become a popular and widely used technique. Unlike the dendrograms resulting from cluster analysis, TWINSPAN tables are ordered and cannot be rearranged like a children's mobile, making them substantially less open to misinterpretation than standard cluster analysis dendrograms.

An example of a simple TWINSPAN table is given in Table 8.6, based on TWINSPAN classification of the Liphook forest fungal dataset. The table lists across the top the ordering of observations (note that the listing starts with observations 15, 22 and 29, signifying plot one in 1988, 1989 and 1990, respectively). Down the left-hand side is the ordered list of species, starting with *Cortinarius semisanguineus*, *Inocybe lacera* and *Lactarius rufus*

Table 8.6 An example of a TWINSPAN classification table, using the Liphook forest fungal dataset. The first table uses the software default values for pseudospecies cut levels (in this case 0, 2, 5, 10 and 20).

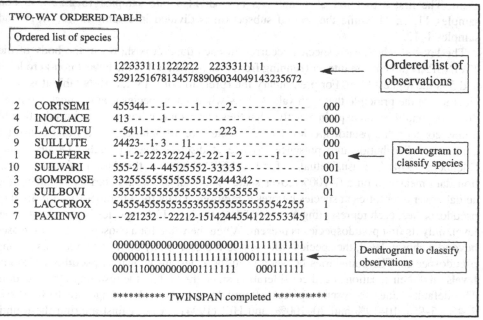

(previously identified as late-stage species) and ending with *Laccaria proxima* and *Paxillus involutus* (early stage fungi).

In addition to the ordering of species and of observation, the table gives three other distinct sets of information. In the body of the table are summarised species frequencies for each observation, while down the right-hand side and across the bottom are dendrograms that divide up the species and the sites respectively.

The dendrograms are very different in appearance to those seen before, but are identical in their information content. They can best be explained by lifting out the top and bottom portions of the TWINSPAN table.

TOP OF TWINSPAN TABLE

```
                  12233311111222222  22333 | 111  1      1
                  52912516781345788906034  | 049143235672
                  ... Body of table ...
                  000000000000000000000001 | 11111111111
the dendrogram    000000111111111111111111 | 000111111111
for samples       000111000000000001111111 | 000111111
                  First dichotomy of samples
```

The first division of the dendrogram is given by the sequence of 0s and 1s in line 1. This shows that samples 15, 22, 29, ..., 34 fall into one division of the data while samples 10, 14, ..., 12 fall into the other. For clarity this first dichotomy is shown by a vertical line in the diagram above (although this is not a standard feature of TWINSPAN tables, which would rapidly become cluttered if many such lines were included). Each of these subdivisions may be further subdivided, as shown by the sequence of 0s and 1s in line 2 of the dendrogram. The first subsection of samples is now divided into samples 15, 22, ..., 35 and samples 11, ..., 34, while the second subsection is divided into samples 10, ..., 19 and samples 1–12.

The level at which each species occurred in each quadrat is shown in the body of the TWINSPAN table by an integer (running from 1 to 5), where 1 is the lowest non-zero level and 5 is the highest level. For preliminary interpretation of a TWINSPAN table it is sufficient to use the principle that high values show where each species is at its most common. To go beyond this and explain exactly what these levels mean involves the introduction of a new concept, the **pseudospecies**. Pseudospecies are artificial constructs that convert a continuous distribution into presence/absence at discrete levels (Mueller-Dombois and Ellenberg, 1974). The usual situation for TWINSPAN involves the classification of vegetation data measured on a 0–100% cover scale. The analysis would start by replacing the actual cover data for each species by presence/absence data for a number (usually five) of pseudospecies, each representing a range of cover values. If the species is present at a low level, only its first pseudospecies is present. When the value for a chosen species increases beyond a chosen level, the species will then be represented by both its first and its second pseudospecies. The boundaries between pseudospecies are known as **pseudospecies cut levels**, and their locations need consideration since they affect the resulting classification. The default values for pseudospecies offered by the original package are (0%), <2%, 2–4%, 5–9%, 10–19% and 20–100%, and Hill (1994) warns against altering these until

Table 8.7 The same data set reanalysed with pseudospecies cut levels set at 0, 1, 5, 10, 20, 50, 100, 500, 1000. Note that the change has altered the details of the dendrograms and the ordering, but has retained very much the same pattern as in Table 8.6.

```
TWO-WAY ORDERED TABLE

                     1221222 11122223233333 111  1       1
                     5298458816701374603125904914 3235672

     2   CORTIN_2    4572--2---------2-344----------- 000
     4   INOCYB_1    423-------2---------------------- 000
     6   LACTAR_1    --5------------223422----------- 000
     9   SUILLUTE    244322---2---------23----------- 000
    10   SUILLU_3    777425 52-4-34665333-2----------- 001
     1   BOLEFERR    --2324 2-222222-2-2--2-2-----2----- 01
     3   GOMPHI_1    332777 7265627653444665542---------- 01
     8   SUILBOVI    785999 93777788777778885555--------- 01
     5   LACCAR_1    945836 39777775555555564877757742556 1
     7   PAXILL_1    --2-22 -232-2222442422255542 265 3345 1

         0000000000000000000000 001111111111111
         0001111111111111111111 10000111111111
             00000000000000111111     000111111
             00001111111111000111
```

experienced with the package. (It is worth being aware that there are several other packages available which include TWINSPAN, and these may offer several standard sets of cut levels.) However if this analysis is extended to use different sorts of data (such as the total counts, as here) one should consider whether pseudospecies cut levels should be redefined. Table 8.8 shows a reanalysis of the Liphook data by TWINSPAN after setting new pseudospecies cut levels, showing that in fact there is minimal change to the ordering of species or observations.

Pseudospecies are usually referred to by their ranked position, so *Laccaria proxima* would generate five pseudospecies: *Laccaria proxima* 1, *Laccaria proxima* 2, ..., *Laccaria proxima* 5. It should be noted that pseudospecies are cumulative, so that if pseudospecies 5 is present (meaning that the actual value is greater than the highest cut level), pseudospecies 1–4 must also be present. Worked examples of this are shown in Table 8.8.

Most users accept the default levels of pseudospecies offered by their package. It is important to check that the cut levels selected are appropriate to your range of data. Clearly the default values of 0–1%, 2–4%, etc. would be quite unsuitable for data which ranges between 0 and 1, but if such an unsuitable set of cut levels were to be chosen no warning would be issued and a classification would be attempted. It is always possible for the user to specify the number of cut levels (between two and nine), and their boundaries. Hill (1994) recommends keeping cut levels low to avoid over-emphasis of dominant species, and warns against setting up too many cut levels since these take up excessive amounts of computer memory and time. The 'front-end' procedure for varying pseudospecies cut levels differs between different implementations of TWINSPAN – as always the message is to check your documentation!

The technical details of TWINSPAN are complicated, and only a bare outline will be given here. More information is given in Kent and Coker (1992) and Hill (1979b, 1994). The basic idea of TWINSPAN is to perform dichotomous divisions of the samples based on the species they contain, then seek species which act as indicators for each of the divisions identified. The technique starts by ordinating the data using correspondence analysis

Table 8.8 The conversion of species data into pseudospecies. The example here refers to the data presented in Table 1.9 and uses pseudospecies cut levels of 0, 2, 5, 10 and 20. It examines the conversion of two species, *Laccaria proxima* and *Paxillus involutus,* into presence/absence data for ten pseudospecies (Lprox1–Lprox5 and Paxinv1–Paxinv5). The results for these species in samples 5, 6 and 12 should be compared with values summarised in Table 8.6, where it may be seen that each species is listed by its highest pseudospecies.

| Observation | Species (with count) | Pseudospecies (presence/absence) | | | | |
	Laccaria proxima	Lprox1	Lprox2	Lprox3	Lprox4	Lprox5
–	0	0	0	0	0	0
5	2	1	1	0	0	0
6	34	1	1	1	1	1
12	99	1	1	1	1	1
	Paxillus involutus	Paxinv1	Paxinv2	Paxinv3	Paxinv4	Paxinv5
–	0	0	0	0	0	0
5	9	1	1	1	0	0
6	7	1	1	1	0	0
12	34	1	1	1	1	1

(Chapter 7), and dividing the first ordination axis at its middle to get a primary division of the samples. This is known as the **primary ordination**. It then searches for species (which are in fact pseudospecies) that are primarily found in just one of the two divisions, known as **indicator species**. The algorithm seeks a limited number of indicator species, typically up to five. It then divides up the data based on the distribution of these indicator species. This second step is known as the **refined ordination**. Generally the results in the final table are based on the refined ordination. However a third step is included to deal with borderline cases. This involves repeating the refined ordination using only a few of the most strongly differential species, and is known as the **indicator ordination**.

The result of these three ordinations is to identify one dichotomous division of the data, and the identification of a set of species which are effective indicators of this dichotomy. This effectively generates two new, smaller datasets which are both independently further subdivided by the same procedure. The divisions terminate when a group is too small (typically five samples) or a predetermined number of divisions have been made (with a default value of six divisions). The TWINSPAN table is then constructed.

The explanatory output generated by a TWINSPAN run gives further insight into the mechanics of the operation. A sample is shown in Table 8.9 from the run on the Liphook fungal dataset which generated Table 8.6. It starts by defining the number of samples and of species. This is followed by defining the numbers given to each species and each sample. (Note that the listing defining sample 1 as observation 1, sample 2 as sample 2, etc. is not as redundant as it may seem. If a sample were missing from the dataset, what the package called sample 39 would be the observation number 40. Failure to realise this could cause serious errors of interpretation.)

Next come listings of the cut levels chosen (0, 2, 5, 10 and 20) and the optional settings. Finally comes the actual classification. Only a small part of this listing is shown – another ten pages were generated by the package. The first ordination (Division 1) is run on the entire 40 observation dataset, and shows axis 1 to have an eigenvalue of 0.298. The first

Table 8.9 Listing generated by TWINSPAN during the preparation of Table 8.6.

************* Two-way Indicator Species Analysis (TWINSPAN) *************
PC-ORD, Version 4.01

Number of samples: 35
Number of species: 10
Length of raw data array: 393 non-zero items

SPECIES NAMES	Part 1: The species names recognised by the procedure
1 BOLEFERR | 2 CORTIN_2 | 3 GOMPHI_1 | 4 INOCYB_1 | 5 LACCAR_1 6 LACTAR_1 | 7 PAXILL_1 | 8 SUILBOVI | 9 SUILLUTE | 10 SUILLU_3	

SAMPLE NAMES

					Part 2: The sample names
1 OBS 01 |	2 OBS 02 |	3 OBS 03 |	4 OBS 04 |	5 OBS 05	
6 OBS 06 |	7 OBS 07 |	8 OBS 08 |	9 OBS 09 |	10 OBS 10	
11 OBS 11 |	12 OBS 12 |	13 OBS 13 |	14 OBS 14 |	15 OBS 15	
16 OBS 16 |	17 OBS 17 |	18 OBS 18 |	19 OBS 19 |	20 OBS 20	
21 OBS 21 |	22 OBS 22 |	23 OBS 23 |	24 OBS 24 |	25 OBS 25	
26 OBS 26 |	27 OBS 27 |	28 OBS 28 |	29 OBS 29 |	30 OBS 30	
31 OBS 31 |	32 OBS 32 |	33 OBS 33 |	34 OBS 34 |	35 OBS 35	

Cut levels:

					Part 3: Pseudospecies cut levels and other options.
0.0000	2.0000	5.0000	10.0000	20.0000	

Options:
  Minimum group size for division = 5
  Maximum number of indicators per division = 5
  Maximum number of species in final table = 200
  Maximum level of divisions = 6

Length of data array after defining pseudospecies: 695
Total number of species and pseudospecies: 47
Number of species: 10
(excluding pseudospecies and ones with no occurrences)

Part 4: Classification of the samples.

CLASSIFICATION OF SAMPLES

****

DIVISION 1 (N = 35) i.e. group*
Eigenvalue: 0.2979 at iteration 3
INDICATORS and their signs:
GOMPHI_1 1(-) SUILVARI 1(-) SUILBOVI 1(-) BOLEFERR 1(-) PAXILL_1 3(+)
Maximum indicator score for negative group − 2
Minimum indicator score for positive group − 1

The first division identifies a group of samples (Obs 1–7 and obs 10) in which *Paxillus involutus* is abundant while *Gomphidius*, *Boletus* and *Suillus* are absent

ITEMS IN NEGATIVE GROUP 2 (N = 23) i.e. group *0

OBS 08	OBS 09	OBS 11	OBS 15	OBS 16	OBS 17	OBS 18	OBS 20
OBS 21	OBS 22	OBS 23	OBS 24	OBS 25	OBS 26	OBS 27	OBS 28
OBS 29	OBS 30	OBS 31	OBS 32	OBS 33	OBS 34	OBS 35	

BORDERLINE NEGATIVES (N = 2)
OBS 08   OBS 09

ITEMS IN POSITIVE GROUP 3 (N = 12) i.e. group *1
OBS 01   OBS 02   OBS 03   OBS 04   OBS 05   OBS 06   OBS 07   OBS 10
OBS 12   OBS 13   OBS 14   OBS 19

NEGATIVE PREFERENTIALS

BOLEFERR1(15, 1)	CORTIN_21(9, 0)	GOMPHI_11(23, 2)	LACTAR_11(7, 0)
SUILBOVI1(23, 3)	SUILLUTE1(9, 0)	SUILLU_31(18, 0)	BOLEFERR2(13, 0)
CORTIN_22(7, 0)	GOMPHI_12(22, 2)	LACTAR_12(5, 0)	SUILBOVI2(23, 3)

(Continued)

Table 8.9 (Continued).

```
SUILLUTE2(6, 0)     SUILLU_32(18, 0)   CORTIN_23(6, 0)    GOMPHI_13(20, 1)
SUILBOVI3(23, 3)    SUILLU_33(15, 0)   CORTIN_24(5, 0)    GOMPHI_14(17, 1)
SUILBOVI4(22, 3)    SUILLU_34(11, 0)   GOMPHI_15(14, 0)   SUILBOVI5(22, 3)
SUILLU_35(8, 0)

POSITIVE PREFERENTIALS
PAXILL_13(5, 9)     PAXILL_14(4, 7)       PAXILL_15(1, 5)

NON-PREFERENTIALS
LACCAR_11(23, 12)   PAXILL_11(18, 12)   LACCAR_12(23, 12)   PAXILL_12(14, 11)
LACCAR_13(23, 11)   LACCAR_14(21, 11)   LACCAR_15(19, 10)

         ------ E N D  O F  L E V E L  1 --------

... + much more of the same!
```

division of the data separates out a group of samples (observations 1–7, 10, 12–14 and 19: these can be identified as representing the earliest successional stage of the community). The package then lists the 23 samples which clearly fell into the negative group, the 12 samples which were clearly in the positive group, and also identifies a small number of borderline samples. It then lists species (actually pseudospecies) according to their preferences. Species tending to occur at the negative end of the ordination axis are called negative preferentials, while those tending to fall at the opposite end of the axis are the positive preferentials. Non-preferentials are species that occur roughly equally in both halves of the ordination axis. This ends the division process for level 1. If this section of the output is compared with the final TWINSPAN table (Table 8.6), it can be seen that samples identified as being in the positive group all lie above the ones that make up the first level of the dendrogram for samples.

TWINSPAN offers the user a number of options to control clustering details and output format, including differential weighting of pseudospecies, number of divisions of the data, and exclusion of rare species from the final table. Most users take default options here until they have gained substantial experience with the package.

TWINSPAN is a widely used and popular approach to the classification of communities, although as yet its use has not widely extended to non-biological data. It has strong similarities to DECORANA, including the point that ease of interpretation has been achieved by a depth of analysis that ensures a high degree of remoteness from the original data! The author himself admits 'TWINSPAN is long and rather complicated' (Hill, 1994). Despite this it is an accepted standard technique which has survived two decades of use without serious criticism. Its main limitation is that it only provides one axis of classification, and may miss information in datasets containing two or more underlying gradients.

9

Canonical correspondence analysis (CCA)

9.1 Introduction

This ordination technique derives from the work of ter Braak (1986), and is frequently used by researchers trying to understand the relationship between community composition and environmental factors. The underlying model is more widely applicable to any situation where the researcher wishes to describe and test the influence of one multivariate dataset upon a second multivariate dataset.

Canonical correspondence analysis is the subject of some nomenclatural confusion, since both it and a related but different ordination technique known as canonical correlation analysis are implemented by a package called CANOCO (ter Braak, 1987, 1990). By convention canonical correspondence analysis takes the initials CCA, canonical correlation is abbreviated to COR, while CANOCO is a computer package (which supplies eight different ordinations). Unfortunately the term CANOCO is loosely used to describe both canonical correspondence analysis and canonical correlation analysis, while CCA has been used to mean COR even in mainstream literature (e.g. Kenkel and Booth, 1992; Matthew *et al.*, 1994).

Like DECORANA and TWINSPAN, canonical correspondence analysis is based on the ordination technique known as correspondence analysis, described in Chapter 7. When a researcher ordinates community data using correspondence analysis (or any of the other ordination techniques previously covered in this book), the resulting patterns often show a strong environmental influence, although interpreting what this actually means may be difficult. The idea of canonical correspondence analysis is to incorporate environmental data into the ordination so as to maximise their importance in the final ordination diagram. In order to perform this analysis two distinct matrices of data are needed, one containing data to be ordinated (the **dependent matrix**, usually population estimates for a number of species) and one describing environmental conditions (the **environmental matrix**) (Figure 9.1). Canonical correspondence analysis uses the same basic algorithm as CA, but the ordination is constrained to fit to a second data matrix containing environmental factors. The second matrix must have the same number of observations (rows) as the species data, but need not have the same number of variables (columns). It is explicitly assumed that the environmental data in the second matrix are paired (in a 1:1 mapping with) the species data. There is a constraint that the number of environmental variables must be fewer than the number of observations (so that the environmental matrix is narrower than it is high).

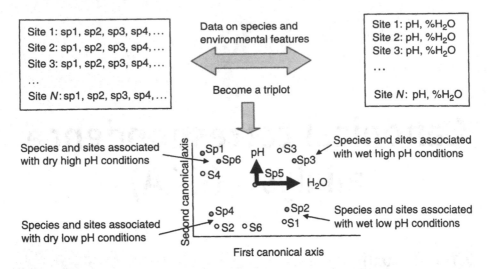

Figure 9.1 The basic idea of canonical correspondence analysis. Dependent data (usually species) are ordinated by correspondence analysis subject to the constraint of fitting to a matrix which defines environmental conditions. This results in an ordination which portrays the major relationships between the two sets of data, and can be portrayed as a triplot summarising the distribution of species, of sites, and of the dominant environmental factors affecting their distribution.

Since canonical correspondence analysis is a modification of correspondence analysis, it inherits the problem of arching distortion when applied to gradient data. This effect has been explained in Section 7.3, and can result in the second ordination axis being merely a distortion of the first axis. For correspondence analysis this was dealt with by the introduction of a detrending algorithm, resulting in detrended correspondence analysis (Chapter 7). Similarly, canonical correspondence analysis can be detrended by the same algorithm giving detrended canonical correspondence analysis (DCCA). This is a little-used technique implemented in the CANOCO package, which is set up and interpreted in the same way as canonical correspondence analysis.

Since environmental data are included from the outset of the ordination, canonical correspondence analysis can be considered to be a form of direct ordination (ter Braak, 1986), although it is so much more complicated than conventional examples of direct ordination that Kent and Coker (1992) describe it as being a hybrid of direct and indirect ordination.

For this technique to be useable it is particularly important to ensure that the two sets of data were collected from the same place and at the same time. If the two sets of data were not collected together, the computer analyses would still run but would be invalid due to the effect known as pseudoreplication. Chapter 1 explains pseudoreplication (and how to avoid it) during the co-collection of biological and environmental data.

The adjective 'canonical' when used in multivariate analysis always has the connotation that a function is being optimised subject to an external constraint. Chapter 10 briefly mentions two other canonical techniques (canonical correlation and canonical variates analysis), which also require a second data matrix to constrain an ordination. Gittins (1985) provides a comprehensive review of this family of techniques.

Table 9.1 A comparison of the algorithms used by correspondence analysis and canonical correspondence analysis.

	Correspondence analysis	Canonical correspondence analysis
step 1	Start with arbitrary (but unequal) site scores	Start with arbitrary (but unequal) site scores
step 2	Calculate species scores as a weighted average of site scores.	Calculate species scores as a weighted average of site scores.
step 3	Calculate new site scores as weighted average of species scores.	Calculate new site scores as weighted average of species scores.
step 4	Standardise site scores.	Perform multiple regression of site scores (weighted by site totals) on environmental variables.
step 5	Stop if site scores are acceptably close to those found in the previous iteration (or an upper limit of the number of iterations has been passed), otherwise return to step 2.	Use multiple regression to derive new, predicted value of site scores.
step 6		Standardise site scores.
step 7		Stop if site scores are acceptably close to those found in the previous iteration (or an upper limit of the number of iterations has been passed), otherwise return to step 2.

9.2 The mechanics of canonical correspondence analysis

The technical details of canonical correspondence analysis are complicated, and the serious researcher is referred to ter Braak (1986, 1987, 1988, 1990) or Jongman, ter Braak and Tongeren (1995) for more mathematical explanations of the procedure. An adequate summary may be given by stating that the standard correspondence analysis ordination is not applied to the species values directly. Instead they are applied to predicted species values derived from a multiple regression of each species in turn on all the environmental variables. The algorithm is iterative, repeatedly recalculating the ordinations and the multiple regressions. The aim of this is to tease apart species' differences in their response to environmental factors.

The difference between correspondence analysis and canonical correspondence analysis is shown by flowcharts describing the algorithms used in both procedures (Table 9.1). It will be seen that the only difference is the addition of two steps which replace actual site scores by values predicted from a multiple regression on environmental data.

A useful image to have here concerns the species response curve to an environmental gradient. The idea of examining species turnover along a gradient was introduced in Chapter 4, under the heading of direct ordination. Consider a hypothetical community of aquatic species, each adapted to a specific set of environmental circumstances. Some will be found in shallow, well-aerated upland streams while others prefer deep muddy rivers lower down in the catchment. It would be possible to plot the occurrence of each species along an axis of one environmental parameter, such as dissolved oxygen levels, and find some separation of species (Figure 9.2a). If information about suspended sediment were added, the species separation could be improved (Figure 9.2b). The idea of canonical correspondence analysis is to find a combination of environmental variables which maximises the species separation (Figure 9.2c).

As would be expected, this technique can produce multiple ordination axes (each explaining progressively less variance), although standard implementations tend to stop at three axes. Each axis contains three distinct sets of information:

(1) **A species ordination**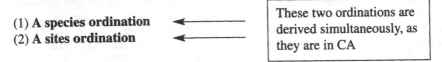

(2) **A sites ordination**

> These two ordinations are derived simultaneously, as they are in CA

(3) **canonical coefficients**, which are weightings defining the importance of each environmental variable to the axis.

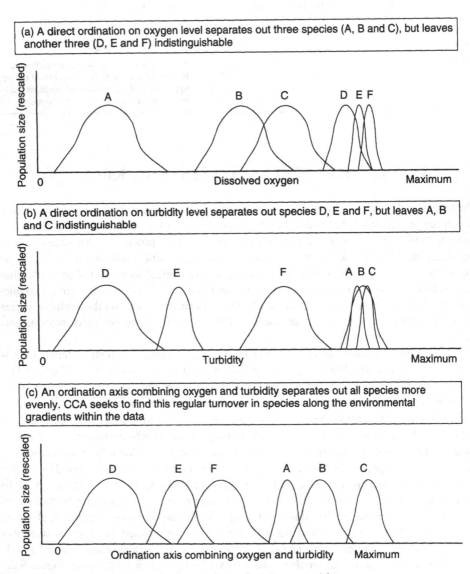

Figure 9.2 Separating species along a gradient of environmental factors.

Remarkably, it is possible to display all three sets of information on the same graph – an extension of the idea of imposing a biplot on ordination diagrams. Such a graph is known as a triplot, an illustration of which is given in Figure 9.1. The ability to produce three sets of analyses simultaneously distinguishes canonical correspondence analysis (along with its sibling technique detrended canonical correspondence analysis) from other ordination techniques. In practice, triplots tend to be too cluttered to be useful interpretive tools. The usual solution is to exclude one set of coordinates, producing a standard biplot. (Up to three such biplots could be produced; species with sites, species with environmental factors, and sites with environmental factors).

The power of canonical correspondence analysis also generates extra difficulties for the researcher, in that the output it generates from a computer package contains an even greater volume of indigestible-looking numbers than other multivariate techniques! Fortunately, the most meaningful insights come from visual inspection of the associated bi/triplots.

9.3 Worked example: CCA on the Wimbledon Common dataset

It would be useful to explain the technique with reference to a simple example in which the environmental/biological relationship may readily be seen. The example used here is the Wimbledon Common dataset (Table 1.4). These two adjoining areas contain very different plant communities and soil types, so would be expected to show clear separation between the environmental associations of the two plant communities. These data were ordinated by canonical correspondence analysis, and the salient sections of the output file are shown in Table 9.2.

Table 9.2 The salient features from a canonical correspondence analysis on the soil/vegetation data of Table 1.4.

1: Introductory information – technique, title of dataset.
********************* Canonical Correspondence Analysis
PC-ORD, Version 2.05
Wimbledon common spoil mound transect

2: Information about options chosen
OPTIONS SELECTED
Axis scores centered and standardized to unit variance
Axes scaled to optimize representation of columns: species
(Scores for species are weighted mean scores for plots)
Scores for graphing plots are linear combinations of species

3: Information about iterations
ITERATION REPORT

Calculating axis 1
Solution reached tolerance of .10E-12 after 14 iterations.

Calculating axis 2
Solution reached tolerance of .10E-12 after 5 iterations.

Calculating axis 3
Solution reached tolerance of .10E-12 after 70 iterations.

(Continued)

Table 9.2 (Continued).

4: Information about the Canonical axes

AXIS SUMMARY STATISTICS
Number of canonical axes: 3
Total variance ('inertia') in the species data: 1.6555

	Axis 1	Axis 2	Axis 3
Eigenvalue	0.992	0.286	0.017
Variance in species data			
% of variance explained	59.9	17.3	1.1
Cumulative % explained	59.9	77.2	78.2
Pearson Correlation, Spp-Envt*	0.997	0.996	0.340
Kendall (Rank) Corr., Spp-Envt	0.929	1.000	0.214

5: Information about contributions of environmental variables to each axis

MULTIPLE REGRESSION RESULTS:
Regression of plots in species space on species

| | Canonical Coefficients | | | | | | |
| | Standardized | | | Original Units | | | |
Variable	Axis 1	Axis 2	Axis 3	Axis 1	Axis 2	Axis 3	S.Dev
1 pH	−0.250	−1.950	0.320	−0.197	−1.538	0.253	0.127E + 01
2 cond	−0.614	1.711	0.073	−0.013	0.037	0.002	0.464E + 02
3 OM	−0.108	0.610	0.229	−0.010	0.054	0.020	0.112E + 02
4 H$_2$O	−0.267	0.854	−0.193	−0.041	0.133	−0.030	0.644E + 01

6: The sites ordination

Scores that are linear combinations of species (LC Scores)
FINAL SCORES and raw data totals (weights) for 8 plots

	Axis 1	Axis 2	Axis 3	Raw Data Totals
1 Heath	0.955005	0.500168	−0.095906	110.0000
2 Heath	0.641382	−0.712307	−0.184144	105.0000
3 Heath	0.797628	−0.076601	0.179300	190.0000
4 Heath	0.844693	0.225777	−0.029678	150.0000
5 Mound	−1.111782	−0.657280	0.176887	86.0000
6 Mound	−1.174268	−0.673791	−0.079795	100.0000
7 Mound	−1.147046	0.424017	−0.135274	93.0000
8 Mound	−1.392911	0.903542	0.059756	94.0000

7: Species ordination. The initials are defined in Table 9.2

FINAL SCORES and raw data totals (weights) for 8 species

	Axis 1	Axis 2	Axis 3	Raw Data Totals
1 Am	−1.158216	−1.589980	3.693880	18.0000
2 Ae	−1.202760	−0.430715	0.101930	265.0000
3 Fr	0.865538	0.671157	0.048975	255.0000
4 Cv	0.755085	−0.884861	−0.793537	240.0000
5 Df	0.874326	0.660151	2.708462	60.0000
6 Hs	−1.309131	2.512848	−0.875099	65.0000
7 Tr	−1.162617	−2.096696	−0.174924	16.0000
8 Vs	−1.235258	1.620400	−2.040935	9.0000

(Continued)

Table 9.2 (Continued).

8: *Correlations between environmental variables and site ordination scores derived from environmental data.*

FINAL SCORES and raw data totals (weights) for 8 species

	Axis 1	Axis 2	Axis 3	Raw Data Totals
1 Am	−1.158216	−1.589980	3.693880	18.0000
2 Ae	−1.202760	−0.430715	0.101930	265.0000
3 Fr	0.865538	0.671157	0.048975	255.0000
4 Cv	0.755085	−0.884861	−0.793537	240.0000
5 Df	0.874326	0.660151	2.708462	60.0000
6 Hs	−1.309131	2.512848	−0.875099	65.0000
7 Tr	−1.162617	−2.096696	−0.174924	16.0000
8 Vs	−1.235258	1.620400	−2.040935	9.0000

9: *Correlations between environmental variables and site ordination scores derived from species data*

INTER-SET CORRELATIONS for 4 species

	Correlations		
Variable	Axis 1	Axis 2	Axis 3
1 pH	−0.978	−0.189	0.009
2 cond	−0.976	0.097	0.013
3 OM	0.902	0.146	0.110
4 H$_2$O	−0.921	−0.104	−0.082

10. *Monte-Carlo test results*

MONTE CARLO TEST RESULTS – EIGENVALUES

	Real data	Randomized data Monte Carlo test, 249 runs			
Axis	Eigenvalue	Mean	Minimum	Maximum	p
1	0.992	0.603	0.171	0.990	0.0040
2	0.286	0.207	0.061	0.300	0.0240
3	0.017	0.079	0.016	0.258	0.9960

p = proportion of randomized runs with eigenvalue greater than or equal to the observed eigenvalue; i.e.,
p = (1 + no. permutations >= observed)/(1 + no. permutations)

MONTE CARLO TEST RESULTS – SPECIES–ENVIRONMENT CORRELATIONS

	Real data	Randomized data Monte Carlo test, 249 runs			
Axis	Spp-Envt Corr.	Mean	Minimum	Maximum	p
1	0.997	0.843	0.484	0.996	0.0040
2	0.996	0.803	0.371	0.996	0.0040
3	0.340	0.578	0.279	0.922	0.9640

p = proportion of randomized runs with species-environment correlation greater than or equal to the observed species-environment correlation; i.e.,
p = (1 + no. permutations >= observed)/(1 + no. permutations)

Strictly it is recommended that the environmental data be checked for collinearity, and that where collinearity occurs it should be controlled by excluding variables. In this case the environmental data are all co-correlated, so strictly only one variable should be entered. We will bypass this step in the procedure in order to highlight some oddities that can occur in CCA when collinear data are entered.

The output has been split up into sections for clarity of explanation – in practice the sections run continuously.

Section 1 starts with information about the technique and package used, and a description for the dataset.

Section 2 defines the options selected during the run. These are technical details about the standardisations and weighting used.

Section 3 gives information about the iterations required for each axis. The algorithm for canonical correspondence analysis involves repeatedly recalculating weights and regressions, and is not guaranteed to converge to a stable solution in an acceptable number of cycles. In this case the difference between cycles (the **tolerance**) was set to 10^{-10}, and was reached in 7, 8 and 9 cycles for axes 1–3 respectively. There will be an upper limit to the number of cycles allowed by the algorithm (20 being a typical value), after which the package will end calculations for the axis and report the actual tolerance achieved. Values above 10^{-5} suggest that the result may be unreliable.

Section 4 gives information about the variance explained by the ordination. As is usual with ordination techniques, the axes decrease in their importance, starting with axis 1 explaining 70.4% of the total variance. There are no firm guidelines for deciding when axes cease to be important, but it seems likely that the third axis (explaining 2.9% of the variance) is random noise. This section also gives species–environment correlations. In this case two different correlation coefficients are used, Pearson's and a non-parametric version of the Kendal rank correlation coefficient. These give the correlation between site scores estimated from species data and site scores estimated as a linear combination of environmental data. The degree of correlation gives an indication of the extent to which environmental factors can explain the distribution of species, but should be handled with caution since even random data can give high values if there are enough environmental variables.

Section 5 gives information about the contribution of environmental data to each axis, in the form of six columns of canonical coefficients. (It also gives the standard deviation for each environmental variable in the final column, under the heading of 'S.Dev'.) It will be recalled that canonical correspondence analysis involves a multiple regression on site scores, using the environmental data as explanatory variables (Table 9.1). This regression gives separate gradients to each of the environmental variables; these are known as the canonical coefficients, but are really just regression coefficients in a multiple regression. As with all multiple regressions, the values of the canonical coefficients can be dangerous to interpret when two or more of the environmental variables are highly correlated (as is often the case). Ter Braak (1986) suggests that where collinearity is a problem, the elegant solution is to exclude one or more environmental variables that are known to be highly correlated with others. Providing the percentage variance explained by the axis changes little, the interpretation of the ordination should not be affected.

In this case it appears that all the environmental variables respond the same way on the first ordination axis, since all have the same (negative) sign. In fact this is not the case, as will be seen below, underlining the dangers in using canonical coefficients to interpret a CCA.

Section 6 gives the results of the ordination of sites. These ordination scores are axis scores, and should be handled in the same way as axis scores in correspondence analysis (Chapter 7). In particular these can be plotted on an ordination diagram (Figure 9.3).

Section 7 gives the species ordination. These again are axis scores, and can be plotted separately or overlain on top of the sites ordination (Figure 9.4). The location of a site within

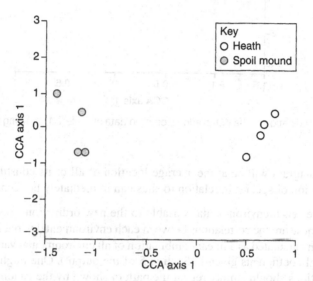

Figure 9.3 CCA ordination of the Wimbledon Common data of Table 1.4, showing the sites ordination.

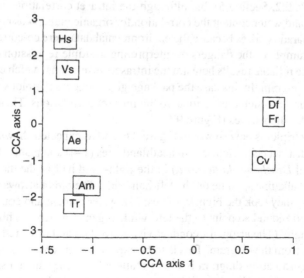

Figure 9.4 CCA ordination of the Wimbledon Common data of Table 1.4, showing the species ordination.

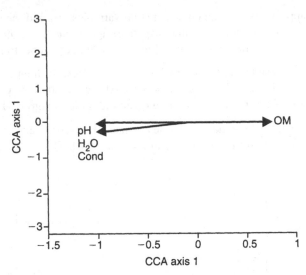

Figure 9.5 CCA ordination of the Wimbledon Common data of Table 1.4, showing the environmental ordination.

the ordination diagram will be at the average location of all of its constituent species, so that the distribution of species in relation to sites can immediately be seen.

Section 8 relates each environmental variable to the new ordination axes, using **intraset correlations**. These are the correlations between each environmental variable and site scores which have been calculated as a linear combination of all environmental variables (weighted by the canonical coefficients given in Section 5 of the output). One might expect that the intraset correlations should simply repeat the pattern shown by the canonical coefficients, but it can be seen that this is not so. All the environmental variables had negative canonical coefficients (Table 9.2, Section 5), but although the intraset correlations were negative for pH, conductivity and water content the correlation for organic matter was positive (Table 9.2, Section 8). This paradox arises because the environmental data were co-correlated (collinear), and is another example of the dangers of interpreting multiple regression coefficients with collinear data. The reliable results here are the intraset correlations, which can be transferred to the ordination diagram. In this case the package gives suggested biplot scores in the right-hand half of Section 8 which are identical to the intraset correlations. These may be plotted as an ordination by themselves (Figure 9.5).

The completed triplot is now shown in Figure 9.6, and as expected shows a very clear separation into two distinct ecosystems, with heathland sites (1–4) and plants (*Calluna vulgaris, Festuca rubra* and *Deschampsia flexuosa*) in the right-hand half of the diagram, while sites and species of the alkaline spoil lie on the left-hand side. The biplot arrows point towards the community where they took the highest values; pH, water content and conductivity were all highest in the spoil mound so point to the left, while organic matter was highest in the heath so points to the right. (The arrow for conductivity is overlain and obscured by that for pH so has been omitted from the diagram. This is to be expected with strongly collinear data.)

The complete ordination diagram is rather cluttered, even for such a small dataset, and would usually be simplified by excluding some information. It is the user's choice which data to omit from the ordination diagram – minor species are commonly left out. In all cases the

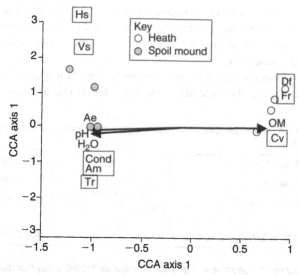

Figure 9.6 CCA ordination of the Wimbledon Common data of Table 1.4, showing the complete triplot.

intention should be to produce a clear and helpful diagram, and the accompanying text should explain what details have been left out.

Section 9 Next the output gives **interset correlations**, which are the correlation coefficients between each environmental variable and site ordination scores derived from species data. These are similar in pattern to intraset correlations.

Section 10: significance testing The significance of the associations between species and environment cannot be gauged from any of the correlation coefficients given above, since the environmental ordination is constrained to correlate optimally with the species data, thereby invalidating the basic assumption made in correlations that the variables may be presumed to be independent of each other. It is possible to perform a formal inferential test of the strength of the association between species and environment, but this can only be found by Monte Carlo methods. This involves the package repeatedly randomly shuffling the data (so that the totals remain the same but any patterns are destroyed) and recalculating the ordination.

It is necessary to consider here what the null hypothesis here is: in fact two are possible. The usual null hypothesis is that there is no association between the matrices. In this case the shuffling operation involves randomly moving rows around in the second matrix (Figure 9.7). At each iteration the eigenvalues defining the strength of the association are recalculated and stored, as are the species–environment correlations. After many iterations the actual distribution of the correlation coefficient for these data builds up, and a dependable probability for the observed value can be found by comparison with the distribution found in randomly allocated data.

The Monte Carlo test results for the null hypothesis of no association are listed in Table 9.2, and examination of the p values listed show the first two axes to be significant while the third is not, both for the eigenvalues and the species–environment correlations. This can be interpreted as suggesting that there is a real, meaningful tendency for the species to be associated with the environmental conditions measured, and that this association may usefully be expressed by examining the first two axes on the CCA ordination. It is important to restate that this does not establish a direction of causality, nor explanation as to why these patterns might have arisen.

An alternative null hypothesis is of no structure in the main matrix. This may be tested by repeatedly randomising the structure of the main matrix (Figure 9.8).

1. First run CCA on the real data, and record eigenvalues and species – environment correlations.

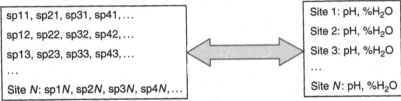

2. Repeatedly rearrange rows in the second matrix and recalculate eigenvalues and correlations. These supply the distribution against which to assess the significance of the value calculated for the source data.

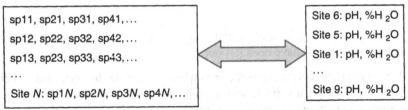

Figure 9.7 The Monte Carlo method to test the null hypothesis of no association between matrices.

1. First run CCA on the real data, and record eigenvalues and species– environment correlations.

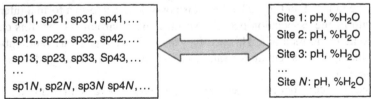

2. Repeatedly rearrange elements within columns within the first matrix and recalculate eigenvalues and correlations. These supply the distribution against which to assess the significance of the value calculated for the source data.

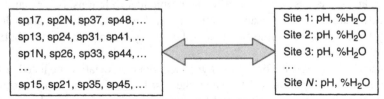

Figure 9.8 The Monte Carlo method to test the null hypothesis of no structure within the main matrix.

To summarise this worked example, canonical correspondence analysis applied to a model dataset has generated a single ordination diagram that neatly encapsulates relationships between all species, sites and environmental factors. In this case the data were sufficiently simple that the trends were clear anyway, and the ordination has highlighted what was already apparent, (multivariate statistics tell you what you already know), but the same approach can be invaluable in the interpretation of larger datasets with less simple divisions.

9.4 Alaskan streams dataset

We can now use CCA to explore relationships between the physical and biotic datasets collected from Alaskan streams (Table 1.11 and 1.12). The species data have been well explored by previous ordinations, but before including the environmental data it is important to check for collinearity in these data. The table of correlation coefficients has been given in Table 1.14, and shows one strong correlation (between conductivity and alkalinity, both being measures of dissolved salts in the water). Following the advice of ter Braak (1990), we will control these by removal of one of these two variables – in this case alkalinity. The issue of data structure again needs to be addressed, due to the potential for pseudoreplication. The biotic data (the dependent variables) consist of total counts of each species for each stream, after transformation by $\log(x + 1)$, and is given in Table 9.3. The environmental data consist of six of the seven properties shown in Table 1.12 (excluding alkalinity, due to its high correlation with conductivity).

The ordination is presented here as the first three axis scores for sites, for species, and for environmental properties (Table 9.4). Figures 9.9 and 9.10 show a simplified version of this, omitting several species plus two environmental factors (CWD or coarse woody debris, and conductivity). The graph of the first two axes (43.4% and 32.9% of the variance explained respectively) identifies two unusual sites, the youngest (Stonefly) and the oldest (Carolus), each with an associated species (*M. insignipes* and *A. guttata* respectively). The environmental factors (shown as arrows) suggest that the youngest site is associated with

Table 9.3 The biotic dataset used in the canonical correspondence analysis of the Alaskan streams dataset.

	stonefly	wolf pt	tyndall	berg n	berg s	rush pt	carolus
Nitocra hibernica	0	0	0	0	0	0	2.838219
Atheyella illinoisensis	0	0	0	0	0	0	1.986772
Atheyella idahoensis	0	0	0	2.285557	1.690196	1.230449	0
Bryocamptus hiemalis	0	0	2.79588	3.25358	1.812913	1.812913	0
Bryocamptus zschokkei	0	0	3.217221	1.690196	2.320146	1.690196	0
Acanthocyclops vernalis	0	3.702517	1.518514	0	1.908485	1.812913	2.247973
Alona guttata	0	0	0	0	0	0	1.690196
Graptoleberis	0	0	0	0	0	0	1.518514
Chydorus	0	0	0	1.518514	0	0	0
Macrothricidae	0	0	0	0	0	0	1.230449
Maraenobiotus insignipes	1.986772	0	1.230449	0	0	0	0

Table 9.4 The CCA ordination of the Alaskan streams data.

	CCA axis		
	1	2	3
% variance	43.4	32.9	16.9
Cumulative%	43.4	76.3	93.2
The sites ordination			
stonefly	−0.86281	4.10017	0.01830
wolf pt	0.33396	−0.19444	−1.99874
tyndall	−0.48972	0.15647	−0.01403
berg n	−0.59483	−0.40319	0.55084
berg s	−0.40406	−0.32708	−0.09007
rush pt	−0.38402	−0.32075	−0.13981
carolus	1.82101	0.13906	0.26352
The species ordination			
nitoh	2.18185	0.21998	0.81247
atill	2.18185	0.21998	0.81247
atida	−0.57880	−0.56791	0.55354
bryhi	−0.58615	−0.33498	0.42580
bryzs	−0.55992	−0.26237	0.15234
bryzs	−0.57797	−0.31963	0.34619
acver	0.33396	−0.19444	−1.99873
along	2.18185	0.21998	0.81247
grapt	2.18185	0.21998	0.81247
chydo	−0.71270	−0.63782	1.69830
Macro	2.18185	0.21998	0.81247
Marin	−0.86281	4.10017	0.01830
The environmental ordination			
pfankuch	0.57841	0.03394	−0.12952
%cwd	0.04363	−0.15136	0.25781
temp	0.24207	−0.35863	0.13687
turbidit	−0.16872	0.66307	−0.15309
conducti	−0.00352	−0.10838	0.15696
age	0.87715	0.02414	0.15539

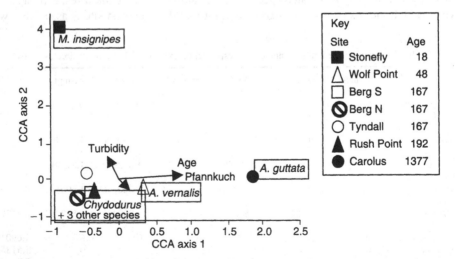

Figure 9.9 CCA ordination of the Alaskan streams dataset, showing first and second axes. (Some observations and species have been removed to reduce clutter.) Species names are italicised in boxes, environmental factors are shown by arrows and labelled by normal text. The unlabelled down-pointing arrow indicates water temperature.

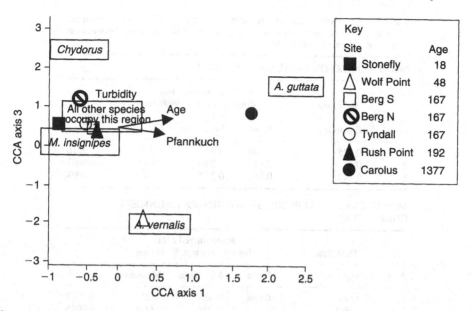

Figure 9.10 CCA ordination of the first and third axes of the Alaskan streams dataset. (Some observations and species have been removed to reduce clutter.)

high turbidity and low temperatures, while the oldest site has the highest age (unsurprisingly) and a high Pfannkuch score. The first canonical axis separates the Carolus community from the others, while the second axis separates out Stonefly.

A similar inspection of the first and third axes (Figure 9.8) shows that the third axis (16.9% of explained variance) isolates Wolf Point as a distinct community, with a particular association with *A. vernalis*.

We can use a Monte Carlo test to estimate the significance of the association. Formally we can test the null hypothesis of no association between the environmental and biological matrices, and the results are listed in Table 9.5, showing clearly that the null hypothesis may be rejected. This shows that there is an association between the matrices, but does not imply the nature of any causal relationship between the datasets.

9.5 Drained mires in Finland

Laine, Vasander and Laiho (1995) studied the vegetation of mires in Finland after drainage for forestry. These mire systems had organic soils that were waterlogged, and contained pines whose growth was limited by the anaerobic soil conditions. It had been noted that drainage led to soil drying and subsidence, associated with increased growth of pines. There were also dramatic changes in the associated vegetation, although the rate at which these changes occurred appeared to differ between sites.

The study recorded vegetation and soil chemical characteristics from 82 sample plots (each 10 × 30 m) on 43 mires, ranging in drainage age from zero (undrained) to 55 years. The patterns found can best be explained by reference to the canonical correspondence analysis performed on the data.

Table 9.5 The Monte Carlo tests produced by running CCA 9999 times on the Alaskan streams dataset, testing the null hypothesis of no association between the biological and environmental datasets.

MONTE CARLO TEST RESULTS – EIGENVALUES

| Axis | Real data Eigenvalue | Randomized data Monte Carlo test, 9998 runs | | | p |
		Mean	Minimum	Maximum	
1	0.825	0.806	0.711	0.825	0.0016
2	0.642	0.542	0.346	0.642	0.0007
3	0.342	0.311	0.227	0.342	0.0008

MONTE CARLO TEST RESULTS – SPECIES–ENVIRONMENT CORRELATIONS

| Axis | Real data Spp-Envt Corr. | Randomized data Monte Carlo test, 9998 runs | | | p |
		Mean	Minimum	Maximum	
1	1.000	0.998	0.989	1.000	0.0037
2	1.000	0.977	0.930	1.000	0.0015
3	1.000	0.968	0.888	1.000	0.0006

Due to the size of the data, the complete biplot (including the species and sites ordination, and the environmental factors) would have been impossibly cluttered. This was solved by several steps. The number of environmental variables included in the analysis was reduced from 31 (measured) to just 10, to avoid the problems of collinearity and to simplify the biplots. The scarcest plant species were also excluded (defined as those found on just one sample plot). Finally, the biplot was presented twice, once to highlight site–environment relationships and once to show species–environment trends. The site–environment biplot (Figure 9.11) shows site numbers (coded by vegetation type and drainage age) overlain with the environmental factors. The site codes are L (low sedge and cotton grass bogs), T (cottongrass – pine fen), V (tall sedge pine fen) and R (herb-rich pine fen). Thus a point labelled R30 indicates a herb-rich fen drained 30 years previously.

A little inspection of Figure 9.11 shows that the first ordination axis corresponds to age since drainage, with the youngest sites at the left of the diagram and the oldest at the right. (This can be seen directly from the overlain arrow for site age, which points approximately parallel to the first axis.) The diagram includes curved, broken lines showing the succession trajectory for each vegetation type. Total tree stand volume and soil carbon content point along the first ordination axis, showing these to increase with site age. This is expected, since the aim of drainage was to assist tree growth, resulting in an increase in leaf litter in the soil.

The second ordination axis can be interpreted as relating to site quality or soil conditions, with the most species-rich communities in the richest soils at the top of the diagram (sites labelled R), and the poorest soils with the slowest plant growth at the bottom (sites labelled T and R). The overlays show that pH and calcium were both highest on the most species-rich sites, since the arrows for these variables pointed up the second axis. Nitrogen and (to a lesser extent calcium) project between the first two axes, showing that these two elements are associated with the better soils, and in addition that they tend to accumulate in the soil with increasing site age.

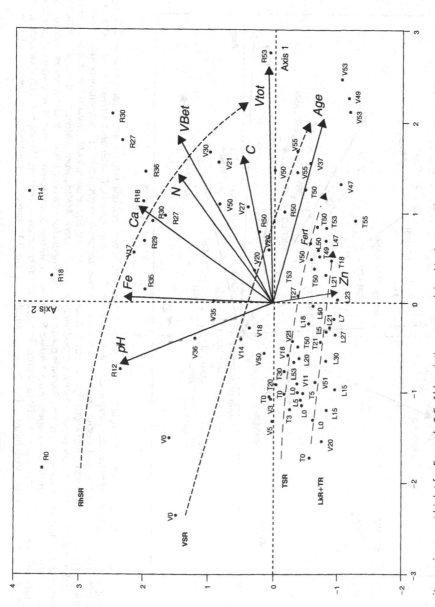

Figure 9.11 The site–environment biplot for Example 9.5. Abbreviations are explained in the text. The environmental variables are: *Age*, drainage age; *Vtot*, total tree stand volume; *VBet*, volume of birches; *pH*, peat pH; *C*, total peat carbon content; *N*, total nitrogen content; *Ca*, total calcium content; *Fe*, total iron content; *Zn*, total zinc content; *Fert*, fertilization, Arrows (dotted lines) added in the diagram show the extent of the vegetation succession in the data for each site type (RhSR = herb-rich pine fen, VSR = tall sedge pine fen, TSR = cottongrass-pine fen, LkR + TR = low sedge and cottongrass bogs). Reproduced by kind permission of the British Ecological Society.

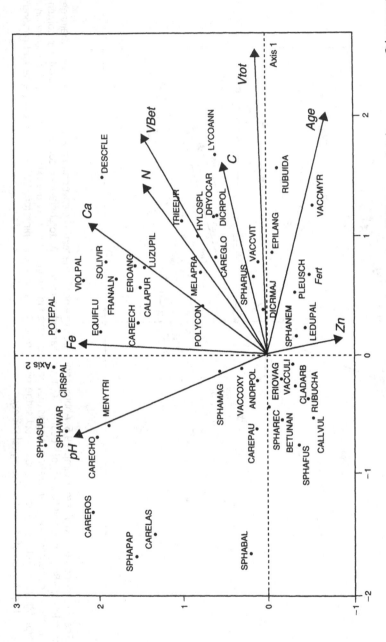

Figure 9.12 The species–environment biplot for Example 9.5. Plant species: ANDRPOL, *Andromeda polifolia*; BETUNAN, *Betula nana*; CALAPUR, *Calamagrostis purpurea*; CALLVUL, *Calluna vulgaris*; CARECHO, *Carex chordorrhiza*; CARECH, *C. echinata*; CAREGLO, *C. globularis*; CARELAS, *C. lasiocarpa*; CAREPAU, *C. pauciflora*; CAREROS, *C. rostrata*; CIRSPAL, *Cirsium palustre*; CLADARB, *Cladina arbuscula*; DESCFLE, *Deschampsia flexuosa*; DICRMAJ, *Dicranum majus*; DICRPOL, *D. polysetum*; DRYOCAR, *Dryopteris carthusiana*; EPILANG, *Epilobium angustifolium*; ERIOANG, *Eriophorum angustifolium*; EQUIFLU, *Equisetum fluviatile*; FRANALN, *Frangula alnus*; HYLOSPL, *Hylocomium splendens*; LEDUPAL, *Ledum palustre*; LUZUPIL, *Luzula pilosa*; LYCOANN, *Lycopodium annotinum*; MELAPRA, *Melampyrum pratense*; MENYTRI, *Menyanthes trifoliata*; PLEUSCH, *Pleurozium schreberi*; POLYCOM, *Polytrichum commune*; POTEPAL, *Potentilla palustris*; RUBUCHA, *Rubus chamaemorus*; RUBUIDA, *R. idaeus*; SOLVIR, *Solidago virgaurea*; SPHABAL, *Sphagnum balticum*; SPHAFUS, *S. fuscum*; SPHAMAG, *S. magellanicum*; SPHANEM, *S. nemoreum*; SPHAPAP, *S. papillosum*; SPHAREC, *S. recurvum complex (S. angustifolium, S. fallax)*; SPHARUS, *S. russowii*; SPHASUB, *S. subsecundum*; SPHAWAR, *S. warnstorfii*; TRIEEUR, *Trientalis europaea*; VACCMYR, *Vaccinium myrtillus*; VACCOXY, *V. oxycoccos*; VACCULI, *V. uliginosum*; VACCVIT, *V. vitis-idaea*; VIOLPAL, *Viola palustris*. Environmental variables as in Fig. 9.11. Reproduced by kind permission of the British Ecological Society.

The species/environment biplot is shown in Figure 9.12, and shows the various species of *Sphagnum* moss to be associated with recently drained sites (being found at the left-hand side of the first ordination axis), while *Lycopodium annotinum* and *Rubus idaeus* were found in the older sites. This fits with the ecology of these species; *Sphagnum* mosses prefer wet conditions while *R. idaeus* is a species of dry woodland soils. *Cirsium palustre* and *Potentilla palustris* preferred the richer more alkaline soils (being at the top of the second axis) while species such as *Calluna vulgaris* and *Vaccinium myrtillus* at the bottom of the second axis preferred acidic infertile soils. (The preferences of the thistle *Cirsium* and ling *Calluna* for good and poor soils respectively reflect long-standing folk knowledge. There is a story of a blind farmer who refused to buy a plot of land because his son could not find a thistle to tie a horse to, showing the soil to be poor. Welsh hill farmers told of 'Gold under gorse, silver under bracken but only bronze under heather'.) The power of the technique is evident in the extent to which all the interpretable information content of this large dataset has been incorporated into two diagrams.

A brief review of other multivariate techniques

This chapter is not intended as a detailed technical guide to any of the named techniques. It is supplied to help the reader understand the basic concepts and applications of the less common multivariate analyses, and to point them to publications where more detailed information may be found. They are listed in alphabetical order.

Canonical correlation analysis (COR) (Gauch and Wentworth, 1976)

This technique is often confused with canonical correspondence analysis (Chapter 9), and is similar in that it requires a second data matrix defining environmental variables in addition to the main matrix (which is usually assumed to contain species data). It dates back to the work of Hotelling (1936), who was interested in knowing how to relate two measures of reading skill in children to two measures of their mathematical ability. This technique may be thought of as a continuation of the progression from bivariate to multiple linear regression. In bivariate linear regression, a best fit is found between one dependent and one independent variable. In multiple linear regression a best fit is found between one dependent variable and a linear combination of many explanatory variables. In canonical correlation analysis a best fit is found between a linear combination of many dependent variables and a linear combination of many explanatory variables. Mathematically the problem can be expressed as follows:

Find A_1–A_j and B_1–B_k so that the canonical correlation R is maximised, where

$$R = \text{correlation between} \quad \sum_{i=1}^{i=j} A_i \times S_i \quad \text{and} \quad \sum_{i=1}^{i=k} B_i \times E_i$$

(10.1)

The species ordination   The environmental ordination

where
A_n is a weighting given to species n
B_n is a weighting given to the nth environmental variable
E_n is the nth environmental variable
j is the number of species

k is the number of environmental variables

S_n is the value for species n.

Notice that in equation 10.1, each paired set of observations $(S_{1,...,j}, E_{1,...,k})$ is reduced to two numbers, one defining the species ordination and one defining the environmental ordination. The correlation between these paired numbers may be calculated in the usual way (the correlation having as many points as there were rows in the data matrices), and is maximised by the COR algorithm. The weightings $A_1–A_j$ and $B_1–B_k$ are known as canonical coefficients and define the canonical axes. There can be as many canonical axes as the number of variables in the smaller dataset (usually the environmental), but all axes are orthogonal and explain progressively decreasing amounts of the variance. As with canonical correspondence analysis, the matrices of species data and of environmental variables must have the same number of rows (observations), though their column numbers will usually differ. There is an important limitation of this technique in that the number of species + environmental variables must be less than the number of sites (i.e. if the matrix of species and of environmental data are joined side by side, the resulting matrix must have more rows than columns). By contrast, CCA (Chapter 9) makes the weaker requirement that the number of environmental variables alone be fewer than the number of observations.

The canonical coefficients are in fact regression coefficients (i.e. gradients) from a multiple linear regression, which means that high levels of collinearity in the data makes their actual values meaningless (see Section 3.2). Jongman et al. (1995) suggest that collinearity should be dealt with by removing selected variables, although Manly (1986) rather forlornly states that when variables are highly correlated there can be no way of disentangling their contributions to canonical variates. Interpretation is by biplot, as in canonical correspondence analysis. Generally these two canonical techniques will find similar patterns within data, and canonical correspondence analysis is to be preferred since it imposes more generous limitations on allowable numbers of variables.

Canonical variates analysis (CVA) (Gittins, 1985; Morrison, 1967)

Otherwise known as canonical discriminant functions (CDF), canonical discriminant analysis (CDA), and multiple discriminant analysis (MDA), CVA is an extension of the two-class discriminant function introduced in Chapter 3. That example used multiple linear regression to allocate individuals to one of two categories by regressing a dummy variable (set to 1 or 2) against their measured properties. CVA attempts to allocate individuals to one from an indefinite number of categories using a linear combination of their measured properties. The aim is to generate a score for each individual which, when entered into a one-way analysis of variance by category, returns the highest possible F value. Since these scores aim to maximise variance within a dataset subject to an external constraint (fitting individuals into categories), the resulting equations are said to be canonical. (This idea is explored further in Chapter 9.) One dataset can generate multiple such canonical equations (up to the limit of the number of categories being fitted minus one, or the number of variables measured, whichever is lower), each being orthogonal and progressively less important than its predecessors. Significance tests are available for each axis, although they assume multivariate normality.

Limitations of this technique are the same as for the two-category model explained under multiple regression. There is an explicit assumption of multivariate normality (Chapter 1), and the problem that the categories need to be known a priori.

Factor analysis (FA)

As explained in Chapter 6, the term 'factor analysis' is used as a synonym for principal components analysis, particularly in the USA. This is not strictly correct, and is to be regretted in that it adds to the confusion surrounding the naming of multivariate techniques. The original meaning of factor analysis comes from a psychometric study by Spearman (1904), who studied correlations between schoolboys' scores in Classics, French, English, Maths, Discrimination of pitch, and Music. The resulting correlation matrix is a classic of multivariate analysis, and is listed in Table 10.1. Spearman suggested that skills in these six different subjects might be explained by a smaller number of hidden but fundamental skills (=factors), whose absolute values could be estimated for each individual.

The formal model for FA involves assuming that a set of p variables $X_1, X_2, X_3, \ldots, X_p$ can be explained by a set of m underlying factors (where $m < p$), where each observation is a linear combination of these factors with a set of weightings, together with a residual error term. Symbolically:

$$X_j = \sum_{j=1}^{j=p} (F_1 \times \lambda_{j1} + F_2 \times \lambda_{j2} + \cdots + F_m \times \lambda_{jm}) + e_j$$

where

F_n is the value of the nth factor
λ_{jn} is the loading of variable j on factor n
e_j is the residual specific to variable j
p is the number of variables
m is the number of factors.

This model allows one to split variance up into the portion attributable to common factors ($\Sigma F_n \times \lambda_{jn}$) – known as the communality – and a remainder (e_j, the specific factors). The factors may be rotated to improve their usefulness, most commonly by the **varimax** rotation which aims to make factor loadings either high or near-zero on each axis (Kaiser, 1958).

It can be shown that FA is an eigenvector problem, and is very similar to PCA in that the eigenanalysis is applied to the correlation matrix. It differs in that the leading diagonal of the matrix has its 1.0s replaced by the highest r value in its column. Factor analysis has a long history in the social sciences, where it is natural to assume that a small number of cognitive skills can explain the great variety of psychometric skills. This is despite the fact that the results are, at best, difficult to interpret and based on assumptions that are probably invalid. Chatfield and Collins (1980) recommend against use of this procedure, while

Table 10.1 Spearman's classic dataset (Spearman, 1904) giving correlations between scores in Classics (C), French (F), English (E), Maths (M), Discrimination of pitch (D), and Music (Mu).

	C	F	E	M	D	Mu
C	1.00					
F	0.83	1.00				
E	0.78	0.67	1.00			
M	0.70	0.67	0.64	1.00		
D	0.66	0.65	0.54	0.45	1.00	
Mu	0.63	0.57	0.51	0.51	0.40	1.00

Hills (1977) went so far as to say that 'FA is not worth the time necessary to understand and perform it'!

Multidimensional scaling (MDS)

This is a generic name for a family of techniques, all of which take a square matrix of distances between individuals and recreates the map from which these distances were measured (Kruskal and Wish, 1978). This is often demonstrated by use of geographical examples, such as recreation of the map of a country given distances between its major cities. As with any ordination diagram, maps generated this way may need to be inverted vertically or horizontally before corresponding to geographical fact. (Of course, maps generated from other types of data need not resemble geographical fact. If multidimensional scaling were applied to political or financial data about countries, the resulting map would show European countries to be far closer to the USA than Cuba, despite Cuba being geographically closer.)

The commonest implementation of MDS uses a matrix of Euclidean distances to recreate a two-dimensional map. This is sometimes referred to as classical multidimensional scaling (CMDS), and is commonly known as principal coordinates analysis (PCO), which is the name used in the original paper by Gower (1966). Chatfield and Collins (1980) suggest the alternative name of classical scaling, although this seems never to have caught on.

Multidimensional scaling is often applied to various forms of environmental data, since it can use matrices containing any valid measure of dissimilarity between individuals, and the resulting 'map' can be used as an informative ordination diagram showing the relative positions of each individual. The commonest distance measure used is Euclidean distance (see Chapters 5 and 8 for a discussion of this and other available distance measures).

Multiresponse permutation procedure (MRPP) (Mielke, 1984)

This is an under-used approach to assessing the probability that two or more groups consisting of multivariate data differ. The groups must be defined a priori. This is conceptually similar to multivariate analysis of variance, but with the important difference that it is non-parametric so may safely be applied to biological/environmental data without worries about multivariate normality. Zimmerman et al. (1985) give an example of this technique being applied to ecological communities to assess differences caused by fires.

Multivariate analysis of covariance (MANCOVA) (Chatfield and Collins, 1980)

This is the multivariate equivalent of analysis of covariance (ANCOVA), in exactly the same way that MANOVA is the multivariate equivalent of ANOVA. It assumes data to be normally distributed, but lacks an equivalent non-parametric test.

Multivariate analysis of variance (MANOVA) (Chatfield and Collins, 1980)

This technique is used when analysing multivariate data arising from a planned experiment, and is directly analogous to univariate ANOVA in that it provides an estimate of the probability that the observed pattern of treatment means would arise in random data. The difference is that each treatment mean is now a coordinate in multivariate space. MANOVA is preferable to performing separate ANOVA tests on each variable in a dataset if the interest lies in assessing whether an overall response has occurred, but will not identify which variables contributed to treatments, should a significant response be detected. This test assumes data to be normally distributed. If normality is questionable (as is often the case for environmental data), MRPP is safer, being an equivalent non-parametric test.

Non-metric multidimensional scaling (NMS or NMDS)

Kruskal and Wish (1978) give a good account of this technique. Also known as ordinal scaling by Chatfield and Collins (1980), this is conceptually similar to multidimensional scaling. A square matrix of distance measures is calculated, and a map is reconstructed. Mather (1976) explains the 'global' version of the algorithm and gives source code, while Prentice (1977) and Minchin (1987) describe the 'local' version. The practical effect of using local or global versions of the algorithm do not seem to have been explored.

Non-metric multidimensional scaling differs in several crucial respects from all other multivariate techniques. Firstly it only uses one distance measure, which is derived from **ranked** differences between individuals. This makes it insensitive to gross deviations from multivariate normality, and hence suitable for data from non-normal, discontinuous or other questionable distributions. This is an advantage for ecological datasets, and this technique is probably under-used.

Secondly, the ordination axes will differ (often dramatically) according to how many axes are requested. The first axis in a one-dimensional ordination is not equal to the first axis in a two-dimensional ordination. Thirdly, where two or more ordination axes are requested *the first axis need not be more important than the second or higher axes*. Axis numbering is arbitrary, totally unlike any other ordination technique.

The choice of how many axes to request is therefore critical but subjective. It is suggested that the user initially requests a high number of axes, and examines how a function called stress (a measure of how well the ordination space corresponds to the original data space) decreases with increasing axis number. The 'correct' number of axes to choose will be the number beyond which stress decreases only slightly. This probably explains the relative unpopularity of the technique.

Procrustes rotation (Digby and Kempton, 1987; Gower, 1975)

This technique is used to compare the results of two different ordinations applied to the same set of data. It holds one set of coordinates fixed, and finds the best fit of the second set of points to this using rotations, rescalings and lateral movements (translations). The effectiveness of the rotation may be assessed by the m^2 statistic, which is the residual sums of squares after the Procrustes operations have been applied. Unfortunately there is no significance test and no clear guidelines to interpret m^2 values. The curious name derives from an inn keeper of Greek mythology who ensured all his customers fitted perfectly to his bed by stretching them or chopping their feet off.

Redundancy analysis (RDA) (Jongman et al., 1995)

This is a derivative of principal components analysis (Chapter 6), with one additional feature. The values entered into the analysis are not the original data, but the best-fit values estimated from a multiple linear regression between each variable in turn and a second matrix of environmental data. Thus the PCA is constrained to optimise a fit to environmental data, so that this technique is the canonical version of PCA (just as canonical correspondence analysis is the canonical version of correspondence analysis, being constrained to correlate optimally with a matrix of environmental data). The output consists of principal axis scores and eigenvectors (as for basic PCA) plus the canonical coefficients arising from the multiple regression. Interpretation is by biplot, and as with all techniques using MLR, collinearity in the data can make the canonical coefficients unreliable. Finally, RDA is noteworthy for being the technique which underlies principal response curves (qv).

Principal response curves (PRC) (van den Brink and ter Braak 1997, 1998, 1999)

This is a new technique, derived from redundancy analysis and intended to facilitate the interpretation of planned experiments applied to biological communities. It assumes that two or more experimental treatments have been applied to a system (usually after a period of pretreatment monitoring), and that one treatment is a control. All treatments are sampled repeatedly to study the temporal development following treatments (perhaps the disappearance and gradual recolonisation of a community following application of a biocide).

The analysis involves applying redundancy analysis to the community data, using experimental treatments to define the matrix of 'environmental' data. This typically generates a rather messy ordination diagram, in which the control treatments meander around randomly while the treated communities diverge rapidly as the treatments take effect.

Principal response curves are then generated by considering each treated plot at each time point and taking the distance to its matched control plot. This converts a tangled ordination diagram into one or more smooth curves, facilitating visual inspection of the responses to the experimental treatments. It is possible to assess the significance of the result by Monte Carlo testing, although careful thought needs to be given to the stage at which randomisation is applied.

Figures 10.1 and 10.2 show the conversion of and RDA ordination into a PRC (Frampton, et al., 2000). The experimental design here involved three irrigation treatments applied to arable soil to study its effect on the community of Collembola (springtails). Samples were taken on ten occasions from -2 days (i.e. pretreatment) to $+98$ days, and the community of animals in each sample used as the dependent data. These counts of animal densities were transformed by $\log(x + 1)$ and subjected to RDA ordination in which the 'environmental' variables were the experimental treatments and the sampling date (each a separate Boolean variable – see Section 1.5.3). The source RDA (Figure 10.1) shows a general trend for later dates to occur in the right-hand side of the ordination diagram, but discerning experimental responses is not easy. Redrawing the graph as a principal response curve shows the hoped-for pattern that treatments overlapped at the start of the experiment (before a treatment had been applied), but diverged later with the greatest response being caused by the added irrigation. The species weights are shown at the side, and together with the PRC itself can predict the response of each species to each treatment at each time point by the simple relationship

$$\frac{(\text{species S in treatment A at time } T)}{(\text{species S in control treatment at time } T)} = \exp(\text{loading}_s \times \text{response}_{At})$$

where
   loading_s is the species loading
   response_{At} is the value of the principal response curve for treatment A at time T (from the graph).

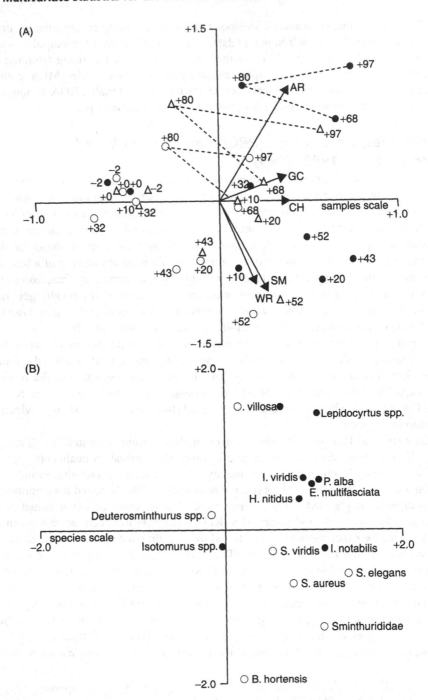

Figure 10.1 An RDA ordination of community data, showing the sites ordination (A) and the species ordination (B). (A) Sample scores for reference rainfall (Δ), spring drought (O) and spring irrigation (●); numbers indicate days relative to start of irrigation treatment. Dashed lines show time-trajectories of treatments from Days +68 to +97. Arrows show the supplementary variables: accumulated rainfall (AR), crop height (CH), ground cover (GC), soil moisture (SM) and rainfall in the week (7 days) preceding sampling (WR). Scale for supplementary variables = 2.0 × samples scale. (B) Species scores for orders Arthropleona (●) and Symphypleona (O).

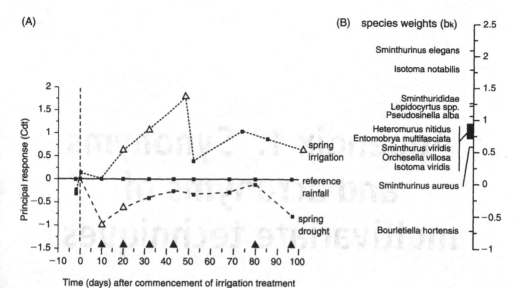

Figure 10.2 The data of Figure 10.1 replotted as a principal response curve. (A) Principal Response Curves (PRC) diagram showing the response over time of a collembolan community to manipulated spring rainfall treatments. (B) Species weights indicating the relative contribution of individual species to the community response displayed in the diagram. Triangles denote significant difference ($p < 0.05$) relative to the reference treatment (△) or between drought and irrigation treatments (▲).

Thus the expected relative abundance of *Sminurinus elegans* in the irrigated plots (compared to the control) at day 43 is $\exp(1.56 \times 2.1) = 26.5$ times higher, while at the same time *B. hortensis* is expected to have $\exp(1.56 \times (-0.74)) = 0.32$ times the density in the same treatment compared to the control.

Appendix 1: Synonyms and acronyms of multivariate techniques

Part 1

This gives an alphabetical list of names of multivariate techniques, indicating the main synonyms and where further information about the technique may be found. Names used in the text are given in **bold**.

Bray–Curtis ordination = polar ordination, Chapter 5.
Canonical correlation analysis (COR), Chapter 10.
Canonical correspondence analysis (CCA), Chapter 9.
Canonical discriminant analysis (CDA) = **canonical variates analysis (CVA)**, Chapter 10.
Canonical discriminant functions (CDF) = **canonical variates analysis (CVA)**, Chapter 10.
Canonical variates analysis (CVA) = canonical discriminant functions = canonical discriminant analysis, Chapter 10.
Classical multidimensional scaling (CMDS), a specific implementation of **multidimensional scaling**.
Classical scaling = **multidimensional scaling (MDS)** = principal coordinates analysis (PCO), Chapter 10.
Cluster analysis = similarity analysis, Chapter 8.
Correspondence analysis (CA) = reciprocal averaging (RA), Chapter 7.
Detrended correspondence analysis (DCA) = DECORANA, Chapter 7.
Detrended canonical correspondence analysis (DCCA), Chapter 9.
Discriminant analysis = **canonical variates analysis**, Chapter 10.
Factor analysis (FA), Chapter 10, but also used to refer to **principal components analysis (PCA)** (Chapter 6).
Information analysis, Chapter 8.
Multiple discriminant analysis (MDA) = **canonical variates analysis (CVA)**, Chapter 10.
Multiple linear regression (MLR), Chapter 3.
Non-metric multidimensional scaling (NMS or **NMDS**), Chapter 10.
Ordinal scaling = **non-metric multidimensional scaling (NMDS)**, Chapter 10.
Polar ordination = **Bray–Curtis ordination**, Chapter 5.

Principal response curves (PRC), Chapter 10.
Principal coordinates analysis (PCO) = classical scaling = (classical) **multidimensional scaling** (C)MDS, Chapter 10.
Principal components analysis (PCA) sometimes referred to as factor analysis, Chapter 6.
Procrustes rotation, Chapter 10.
Reciprocal averaging (RA) = **correspondence analysis (CA)**, Chapter 7.
Redundancy analysis (RDA), Chapter 10.
Scaling – a generic term for producing maps based on distances between individuals.
Similarity analysis = **cluster analysis**, Chapter 8.
Two-way indicator species analysis (TWINSPAN), Chapter 8.

Part 2: Standard acronyms for the major multivariate techniques

CA = correspondence analysis.
CANOCO = A computer package offering CCA, COR, DCCA and other techniques (Ter Braak 1987, 1990).
CCA = canonical correspondence analysis.
CDA = canonical discriminant analysis.
CDF = canonical discriminant functions.
CMDS = classical multidimensional scaling.
COR = canonical correlation analysis.
CVA = canonical variates analysis.
DCA = detrended correspondence analysis.
DCCA = detrended canonical correspondence analysis.
DECORANA = A computer package implementing RA and DCA (Hill, 1994).
FA = factor analysis.
MANCOVA = multivariate analysis of covariance.
MANOVA = multivariate analysis of variance.
MDA = multiple discriminant analysis.
MDS = multidimensional scaling.
MLR = multiple linear regression.
NMDS = non-metric multidimensional scaling.
NMS = non-metric multidimensional scaling.
PCA = principal components analysis.
PCO = principal coordinates analysis.
PRC = principal response curves.
RA = reciprocal averaging.
RDA = redundancy analysis.
TWINSPAN = two-way indicator species analysis.

Appendix 2: A general approach to linear regression using linear algebra

2A1 Introduction

It is possible to derive bivariate and multiple regression equations, perform a significance test comparing models, and even extend the analysis to include covariance Y, all based on one general model known as the **general linear model**. This model requires that the data be considered as matrices, and is written in simple matrix algebra. It is not essential to follow the contents of this appendix in order to perform and use simple or multiple regression, but comprehension of its principles will deepen the reader's understanding. In order to follow this section the reader will need to be conversant with matrix multiplication, addition, and the process of Gaussian elimination.

The structure of this appendix is that the general linear model will be introduced first for bivariate regression using a worked example. This model will be extended to general multiple regression, and finally significance testing using analysis of variance tables will be introduced.

2A2 Bivariate regression

The starting point is to rewrite the standard bivariate regression equation (equation 3.1) to include the residuals:

$$Y_i = A + B \times X_i + e_i \qquad (3.9)$$

where
  Y_i is the ith value of the dependent variable
  A is the intercept
  B is the regression coefficient (or gradient of Y on X)
  X_i is the ith value of the independent variable
  e_i is the ith residual.

The constraint is to find values of A and B which minimise the value of the sum of residuals squared ($\sum e_i \times e_i$).

This equation may now profitably be rewritten in matrix algebra, in which each letter now represents a rectangular array of numbers:

$$\mathbf{Y} = \mathbf{X} \times \boldsymbol{\beta} + \boldsymbol{\varepsilon} \qquad (2A.1)$$

where

$\mathbf{Y}$ is a $1 \times N$ matrix containing only values of the dependent variable

$\mathbf{X}$ is a $2 \times N$ matrix, of which the first column consists of 1s and the second column contains the independent variable

$\boldsymbol{\beta}$ is a 1×2 matrix containing values of A and B (the intercept and gradient respectively)

$\boldsymbol{\varepsilon}$ is a $1 \times N$ matrix containing the residuals.

This model can be illustrated with the data and regression equation shown in Figure 3.9 (FDA-measured enzyme activity against respiration rates for nine plant litters). The values for X, Y and the residuals are all shown in Figure 3.9. It can be verified by hand that the values are consistent, as follows:

$$40.28 = (1 \times 0.7733) + (136.8 \times 0.2256) + 8.648$$

$$14.46 = (1 \times 0.7733) + (72 \times 0.2256) - 2.5546$$

etc.

Data from Figure 3.9 replotted as the components of a linear model:

Y	=	X		×	β	+	ε
40.28		1	136.8				8.6482
14.46		1	72				−2.5546
13.73		1	68.4				−2.4726
8.98		1	41.4		0.773		−1.1321
13.73	=	1	91.8	×	0.226	+	−7.7510
31.17		1	115.2				4.4106
23.4		1	82.8				3.9492
27.94		1	161				−9.1507
27.94		1	93.6				6.0530

The elements of $\boldsymbol{\beta}$ give the parameters of the best-fit regression equation:

$Y = 0.773 + 0.226 \times X$

This format does not yet tell us how to calculate the elements of $\boldsymbol{\beta}$, but provides a full formal definition of the linear model. It is now possible to derive the sum of residuals squared as $\boldsymbol{\varepsilon}'\boldsymbol{\varepsilon}$ (where $\boldsymbol{\varepsilon}'$ means the transpose of matrix $\boldsymbol{\varepsilon}$), and from this to show that $\boldsymbol{\beta}$ may be found as follows:

$$\mathbf{X}'\mathbf{X}\boldsymbol{\beta} = \mathbf{X}'\mathbf{Y} \qquad (2A.2)$$

This equation may be solved by matrix inversion:

$$\beta = (X'X)^{-1} X'Y \qquad (2A.3)$$

It is a good exercise for the reader to use this matrix inversion relationship to derive the regression equation for the standard Y–X regression given in Figure 3.6, but for technical reasons to do with computational precision, matrix inversion is not implemented inside analytical packages. Instead equation 2A.2 is usually solved inside computers by the process of Gaussian elimination (essentially the solving of simultaneous equations). The reader is referred to standard texts on linear algebra for further explanation and suitable algorithms.

2A3 Multiple linear regression

Exactly the same process can be used to fit two independent variables X and Z to a dependent variable Y: the only difference now is that the matrix **X** now contains three columns (a column of 1s, a column of the first independent variable and a column of the second explanatory variable) while the matrix β contains three elements (a constant, the gradient of Y with respect to X and the gradient of Y with respect to Z).

The general linear model for a regression with two explanatory variables X and Z

$$Y = X \times \beta + \varepsilon$$

is solved by

$$X'X\beta = X'Y \qquad (2A.1)$$

Where the matrices contain the following:

Y	=	X			×	β	+	ε

Y		=	X			×	β	+	ε
Y1			1	X1	Z1		A		residual 1
Y2	=		1	X2	Z2	×	B	+	residual 2
Y3			1	X3	Z3		C		residual 3
Y4			1	X4	Z4				residual 4

Having solved β, its elements give the parameters of the best-fit regression equation:

$$Y = A + B \times X + C \times Z$$

These elements are found by precisely the same Gaussian elimination process as described above. The procedure generalises, so that three, four or more explanatory variables may be fitted. The key element to consider is the composition of the columns in the matrix **X**. The format of this matrix defines the relationship being assumed between **Y** and the explanatory variable(s). This model is known as the general linear model, and it lies at the heart of many analyses and modelling packages. More complex models may be fitted by modifications to the **X** matrix: an example of fitting a polynomial function was given in Table 3.3. Two separate regression lines may be fitted within one graph by allowing an independent variable to occupy two columns within the matrix **X**, with values either being entered or left as zero.

The general linear model for a regression with one explanatory variable X, but fitting separate lines for two subsets of the data (perhaps two sites, or two species):

$$\mathbf{Y} = \mathbf{X} \times \boldsymbol{\beta} + \boldsymbol{\varepsilon}$$

So

$$\mathbf{X}'\mathbf{X}\boldsymbol{\beta} = \mathbf{X}'\mathbf{Y}$$

Where the matrices contain the following:

$\mathbf{Y}$	$=$		$\mathbf{X}$		$\times$	$\boldsymbol{\beta}$	$+$	$\boldsymbol{\varepsilon}$
Y1		1	X1	0				residual 1
Y2	$=$	1	X2	0	$\times$	A	$+$	residual 2
Y3		1	0	X3		B		residual 3
Y4		1	0	X4		C		residual 4

Here A is the constant, B is the gradient of Y with respect to X for the first two observations, and C is the gradient of Y with respect to X for the second two observations.

2A4 Significance testing

It is possible to extend the framework of the general linear model to incorporate formal significance testing, using analysis of variance. This relies on the fact that the sum of residuals squared ($= \text{residual}_1^2 + \text{residual}_2^2 + \cdots + \text{residual}_n^2 = \boldsymbol{\varepsilon}'\boldsymbol{\varepsilon}$) can be assumed to behave in the same way as summed deviations from a mean – known as a 'sum of squares'. The key point is to define the degrees of freedom associated with this sum of squares, which is the total number of observations minus the number of elements in the matrix $\boldsymbol{\beta}$.

The null model is that the data have no relationship:

$$Y = \text{constant} + \text{error}$$

This is defined by a linear model in which the matrix $\mathbf{X}$ consists solely of 1s. This will solve the least-squares best fit as

$$Y = \text{mean}(X)$$

The sum of squares is now $\Sigma(X - \text{mean}(X))^2$ with $n - 1$ degrees of freedom. Call this SS_1.

The simplest alternative model is a linear relationship between Y and X:

$$Y = \text{constant} + B \times X + \text{error}$$

This is defined by a two-column matrix $\mathbf{X}$, and after solving the general linear model a new fit will be obtained with a smaller sum of squares that has $n - 2$ degrees of freedom. Call this SS_2. The significance of the model can be examined by standard analysis of variance, testing the null hypothesis that the residuals from the regression line come from the same population as the residuals from the mean.

Analysis of variance table for a bivariate regression. SS_1 and SS_2 are defined above.

Source	DF	Sum of squares	Variance or mean square	Variance ratio F
Regression	1	$SS_1 - SS_2$	$SS_1 - SS_2$	by division
Error	$n - 2$	SS_2	$SS_2/(n - 2)$	
Total	$n - 1$	SS_1		

Again it is a good exercise for the reader to use this approach to show that there is a simple relationship between the variance ratio F and the Pearson correlation coefficient r:

$$F_{1,n-2} = (n-2) \times \sqrt{\frac{r^2}{1-r^2}} \tag{2A.4}$$

Each progressive increase in the complexity of a linear model can be tested by this approach: a change in the model changes the degrees of freedom and the sum of squares – a change which can then be tested by ANOVA. This approach is at the core of stepwise regression models, described above, in which variables are progressively added until there is no significant improvement in the model.

The approach also allows us to see why a zero-intercept model (Figure 3.7) may not be subjected to an inferential test. This is because it cannot be built up from the null model Y = constant (which has $n - 1$ degrees of freedom). The zero-intercept model is based on the relationship Y = constant $\times$ X, which also has $n - 1$ degrees of freedom, so the ANOVA model allows $1 - 1$ = zero degrees of freedom to test this model.

Finally, note that the reduced major axis approach to calculating a best-fit line (Figure 3.8) is simply not amenable to the general linear model because the residuals are not vertical, a requirement which is implicit in any equation based on the general format:

$$Y = \text{(some calculation)} + \text{residual}$$

The appropriate mathematics for this family of models are touched on in Chapter 6.

Appendix 3: Multivariate software for the environmental researcher

3A1 Preface

This section aims to list the name, suppliers and features of the more commonly used software packages currently available which might be of use to researchers. It does not aim to promote or support any of these, and the final choice must remain dependent on the user's judgement and budget. A failure to include a package here should not be taken to imply that any software is unsuitable, and the author would welcome being informed of any updates or unlisted software. The analyses listed in this book are itemised as separate columns in Table 3A3, and each package is defined by its capabilities under these headings. Specifications change rapidly and it is essential that the user confirms details from the manufacturer before purchasing software. The general packages listed here could run some multivariate analyses additional to those listed here by use of 'plug-ins' or macros.

There are many useful Internet pages about ordination techniques, though the links will inevitably change over time. At present there is an excellent page linking to these, plus other useful software hosted by Oklahoma University called Ordination Methods for Ecologists at http://www.okstate.edu/artsci/botany/ordinate/

3A2 The software

ADE-4: Downloadable software available as a series of modules, currently from http://pbil.univ-lyon1.fr/ADE-4/ADE-4.html

CANOCO/CANODRAW: CANOCO is a specialist ordination package written by Ter Braak, based around canonical techniques (CCA, DCA, DCCA). CANODRAW is the accompanying software for drawing ordination diagrams. Produced by Microcomputer Power, 111 Clover lane, Ithaca, New York 14850, USA.
URL: http://www.microcomputerpower.com/

Community Analysis Package: Pisces Conservation LTD, IRC House, The Square, Pennington, Lymington, Hants., SO41 8GN. United Kingdom.
URL: http://www.pisces-conservation.com/

CLUSTAN: Primarily intended for cluster analysis, written by David Wishart.
URL: http://www.clustan.com/

DECORANA: A modern version of the original program by Mark Hill of ITE. Comes with TWINSPAN, also TABLCORN, software for converting data between Cornell Condensed Format (Chapter 7) and other formats. Supplied by CEH, the descendants of ITE.
URL: http://www.ceh.ac.uk/products_services/software/corn.html

GENSTAT: A general data-handling package, which includes extensive data manipulation facilities and many univariate techniques in addition to multivariate analyses. This package has long associations with The Lawes Agricultural Trust, Rothamsted.
URL: http://www.nag.co.uk/stats/tt_soft.asp

MVSP: A general multivariate analysis package, authored by Warren Kovach.
URL: http://www.kovcomp.co.uk/

PC-ORD: Primarily intended for ordination of ecological data. Access to Microsoft EXCEL is helpful. MjM Software, Gleneden beach, Oregon USA.
http://www.ptinet.net/~mjm

PRIMER: A package based on non-parametric analyses and permutation-based tests, originally developed by Plymouth Marine Laboratories for analysing the impact of pollution on aquatic systems, but now a powerful general package sold by a separate company: PRIMER-E Ltd, 6 Hedingham Gardens, Roborough, Plymouth PL6 7DX United Kingdom.
URL: http://www.pml.ac.uk/primer/index.htm.

SAS: A general data-handling package, which includes extensive data manipulation facilities and many univariate techniques in addition to multivariate analyses.
URL: http://www.sas.com/

SPLUS: A general data-handling package, which includes extensive data manipulation facilities and many univariate techniques in addition to multivariate analyses. Produced by Mathsoft.
URL: http://www.mathsoft.com/

SPSS: A general data-handling package, which includes extensive data manipulation facilities and many univariate techniques in addition to multivariate analyses.
Produced by SPSS inc.
URL: http://www.spss.com/

Statistica: A general data-handling package, which includes extensive data manipulation facilities and many univariate techniques in addition to multivariate analyses. Produced by Statsoft inc.
URL: http://www.statsoftinc.com/

Table 3A3 Summary of capabilities.

Name	Diversity	MLR	Bray–Curtis	PCA	CA	DCA	Cluster analysis	TWINSPAN	CCA + MC	Other?
ADE-4	–	–	–	Yes	Yes	–	–	–	Yes	CVA
CANOCO/ CANODRAW	–	–	–	Yes	Yes	Yes	–	–	Yes	RDA, DCCA
Community Analysis Package	Yes	Yes	–	Yes	Yes	Yes	Yes	Yes	Yes	NMDS, association analysis, species filtering[1]
CLUSTAN	–	–	–	Yes	–	–	Yes	–	–	Many forms of clustering including K-means
DECORANA/ TWINSPAN/TABLCORN	–	–	–	Yes	Yes	Yes	–	Yes	–	–
GENSTAT	–	Yes	–	Yes	Yes	–	Yes	–	–	MDS, procrustes rotation, CVA
MVSP[2]	Yes	–	–	Yes	Yes	Yes	Yes	–	Yes	MDS
PC-ORD	Yes	–	Yes	Yes	Yes	Yes	Yes	Yes	Yes[3]	Jack-knifing, NMS, MRPP MDS, ANOSIM, SIMPER[4]
PRIMER-E	Yes	Yes	–	Yes	–	–	Yes	–	–	–
SAS	–	Yes	–	Yes	–	–	Yes	–	–	CVA, MDS, COR
SPLUS	–	Yes	–	Yes	–	–	Yes	–	–	CVA, MDS, COR
SPSS	–	Yes	–	Yes[5]	–[6]	–	Yes K-means	–	–	CVA, FA, MDS, K-means clustering
Statistica	–	Yes	–	Yes	Yes	–	Yes	–	–	NMDS, FA, CVA, MDS, COR

1. Novel method producing a two-dimensional ordination of sites and species.
2. There is also a plug-in for Microsoft EXCEL.
3. The package offers a similar species–environment ordination to CCA, called Bio-Env, with Monte Carlo testing.
4. These are non-parametric versions of discriminant functions and multivariate ANOVA.
5. PCA is listed under 'factor analysis'.
6. The package offers correspondence analysis, but only as an option to analyse nominal data in a contingency table, and as such cannot be used to ordinate continuous multivariate data.

Appendix 4: Suggested further reading

Specifically on collection and analysis of floral data (with good explanations of ordination techniques):

Kent, M. and Coker, P. (1995).
Vegetation Description and Analysis: A Practical Approach.
Wiley.

Specifically about ecological diversity:

Magurran, A.E. (1988).
Ecological Diversity and Its Measurement.
Chapman & Hall.

More advanced reading on multivariate analysis:

Jongman, R.H., Ter Braak, C.J.F. and Van Tongeren, O.F.R. (1995).
Data Analysis in Community and Landscape Ecology.
Cambridge University Press.

Hair, J.F., Anderson, R.E. and Tatham, R.L. (1998).
Multivariate Data Analysis.
Prentice Hall.

Johnson, R.A., Wichern, D.A. and Wichern, D.W. (1998).
Applied Multivariate Statistical Analysis.
Prentice Hall.

Glossary

agglomerative: an adjective referring to the forms of classification analysis in which individual observations are progressively linked together to form larger clusters (Figure 8.1). Its opposite is 'divisive', covering classification techniques which start with the entire dataset and progressively divide it down.

analysis of covariance: a model which combines features of regression and ANOVA to test for differences between group means while controlling for the bias introduced by a known correlation with another measured factor. As with ANOVA, this can build up into complex multilayered models.

analysis of variance: a univariate inferential technique which tests the null hypothesis that there are no differences between group means (where each group is assumed to represent a treatment). ANOVA is most commonly run to test the effects of one treatment (one-way ANOVA) or two combined groupings (two-way ANOVA) but can build up into complex multilayered models, given a suitable design. The reader is referred to any standard statistical textbook.

ANCOVA: a standard acronym for analysis of covariance (qv).

ANOVA: standard acronym for analysis of variance (qv).

axis: a mathematical line used to plot a graph. Axes can be measured data (as in a standard scattergraph, such as Figure 3.1), but in this book the term 'axis' often refers to an ordination axis, which is a compound axis derived from many variables in order to facilitate data examination.

CANOCO: a computer package written by ter Braak (1987) which performs a variety of ordination techniques including CCA, RDA and DCCA.

canonical: an adjective referring to an ordination which has been constrained to correlate optimally with a second dataset – a constrained ordination. The regression coefficients produced in the process are known as canonical coefficients.

Czekanowski similarity coefficient: an index of similarity between two samples which may be used as the starting point in Bray–Curtis ordination, cluster analysis or other distance-based ordinations. It is defined in Table 5.1.

city block metric: a distance measure, defined in Table 8.4, which can be used to provide a metric of distances between pairs of observations that may be used as an input to cluster analysis or certain ordinations (PCO, Bray–Curtis).

cluster analysis: a large topic (covered in Section 8.3) which seeks to classify data into clusters of similar observations.

collinear, collinearity: this describes datasets where some of the variables used as explanatory are themselves co-correlated. Under these circumstances the regression coefficients (the gradients of the best-fit line) may be misleading, although the correlation coefficient remains correct.

complementary analysis: the overlaying of cluster-analysis dendrograms onto ordination diagrams. An example is given in Figure 8.11.

constant (of a regression): see intercept.

constellation diagram: a visual portrayal of patterns of association between species in a set of communities.

continuous data: data whose values are expressed with (potentially unlimited) decimal places: height, mass, pH, etc. See also ordinal and nominal.

correlation coefficient: an index of the extent to which two variables change in value together, usually expressed by Pearson's product moment correlation coefficient (r), which ranges from 1 to -1 (Figure 3.13). R^2 is equal to the proportion of the variance which is explained by the linear regression.

correlation matrix: a square matrix whose rows and columns relate to the same set of variables, and for whom the (I,J)th value would be the correlation between variable I and variable J. An example of a correlation matrix is given in Table 6.3.

DECORANA: a FORTRAN program (described in Section 7.3) to run a detrended correspondence analysis on ecological data.

dendrogram: a tree-like branching diagram showing relationships between data points. Examples are given in Figures 8.12 and following.

dependent variable: the variable which is assumed to be dependent on other variables, always plotted on the Y-axis. A dependent variable must be identified in any regression model, but the concept does not usually feature in multivariate analyses.

dissimilarity: another name for a distance between observations.

dissimilarity matrix: see distance matrix.

distance matrix: a square matrix, similar to a correlation matrix, in which the rows and column define the same list of observations, but here the (I,J)th value is the distance between observations I and J using one of a variety of distance functions. An example is given in Table 5.3.

diversity index: a numeric score given to a community which reflects its diversity. There is no simple definition of diversity – see Chapter 2.

divisive: (in the sense of classification techniques) any technique which starts off with the entire data structure and progressively classifies it by breaking it down into smaller units (the opposite of agglomerative – qv).

eigenvalue: in pure mathematics, eigenvalues are properties of a matrix when it is repeatedly used to multiply another matrix. It will be found that a stable pattern emerges which grows or shrinks at a constant relative rate (see Figure 6.8) – the eigenvalue is the rate of growth of the equilibrium pattern. In multivariate analyses eigenvalues indicate the strength of a pattern within an ordination, and as such indicate the percentage variance explained by a given ordination axis.

eigenvector: in pure mathematics, eigenvectors are properties of a matrix when it is repeatedly used to multiply another matrix. It will be found that a stable pattern emerges which grows or shrinks at a constant relative rate (see Figure 6.8): this pattern is the first eigenvector, and if its influence is removed, second and higher (weaker) eigenvectors will emerge. In multivariate analyses an eigenvector often describes the relative contributions of variables to an ordination axis, and as such can be used to interpret patterns in a user's data.

Euclidean distance: a distance measure (qv) in which distances between points in data space are measured exactly as it would be with stretched string – the familiar distance measure used in our real space.

explanatory variable: one of the variables in a regression model which is assumed to be affecting the behaviour of the dependent variable (qv). Although there can usually only be one dependent variable there may be many explanatory variables in a regression model. An alternative name (conventional for bivariate models) is the independent variable, but this is not encouraged in multiple regression models since these 'independent' variables are very often not independent of each other.

Gaussian distribution: see normal distribution.

heteroskedastic, heteroskedasticity: the tendency of the variance of a distribution to increase as values become larger (Figure 3.4), the opposite of homoskedasticity.

homoskedastic, homoskedasticity: of a distribution, when its variance remains approximately constant over a range of values. The opposite of heteroskedasticity.

independent variable: see explanatory variable.

inferential test: a test which returns a probability by which a user may judge whether or not to accept a null hypothesis. This includes familiar tests such as the t test, ANOVA and correlations, but excludes many multivariate techniques.

intercept: the point at which a fitted line passes through the Y-axis, the value of Y when all explanatory variables are zero (see Figure 3.5).

interval data: data which are continuous (qv) but for which the zero is arbitrary: addition and subtraction are meaningful but multiplication or division are not. (The only common example is temperature: 30°C is 10°C higher than 20°C but not three times hotter.)

Jaccard similarity coefficient: a distance function (qv) defined in Table 5.1.

jack-knifing: a computer-intensive technique in which the distribution of a community index (a regression coefficient or a diversity index) is estimated by excluding one observation at a time and recalculating the index. Thus a dataset with N observations can generate N different 'pseudovalues' of the index. This is covered in Section 2.4.

K-means clustering: a clustering technique described in Section 8.1, which uses random starting points.

least square: an analytical approach in which the best model is selected on the basis of minimising the sum of all residual values squared. This is the standard criterion for assuming a model to be optimum.

logarithmic transformation: the replacement of a variable by a new one calculated as the logarithm of the source data. In the case of data containing zeros (which includes all species data) the transformation used is $\log(X + 1)$, thereby transforming 0 into 0 rather than an illegal minus infinity. This is commonly done to remove heteroskedasticity (qv) from data or to smooth down the influence of outliers.

mean: the correct mathematical term for an average: add up the values and divide by the number of observations. This is in fact the least squares solution to a model which holds all observations to have the same value, but with added random noise.

Monte Carlo testing: a general term for any significance test in which the distribution of a calculated value (such as a correlation coefficient or an eigenvalue) is estimated by repeatedly imposing a random shuffling on a dataset and recalculating the value. This is a computer-intensive process, but for all but very small datasets is a safe and very general technique for estimating significance. Different patterns of shuffling allow the testing of different hypotheses.

monothetic: a little-used term describing techniques for divisive (qv) classification of ecological communities in which each division is based on the counts of just one species. Its opposite is polythetic – most classification techniques are polythetic.

nominal data: data which consist of counts within named categories (Section 1.5.1). There are rather limited options available for multivariate analysis of nominal data, except for those techniques which start with a matrix of distances between observations, since distance functions (qv) are available for nominal data (see Table 5.1).

normal distribution: the normal or Gaussian distribution (after Karl Gauss who solved the equation) is the expected distribution when large numbers of random distributions are superimposed. It gives rise to the famous 'bell curve', and its presence in data is assumed by all parametric inferential statistical tests. Formally, to define a normal distribution one only needs to know its mean M and standard deviation S.

ordinal data: data which can be placed into ascending order, but not added or subtracted (Section 1.5.1).

ordination: the operation of arranging observations along in a defined order. This order may be given by measurements on some preknown environmental data (direct ordination) or along an abstract mathematical axis (indirect ordination).

Pearson's correlation coefficient (correctly Pearson's product moment correlation coefficient): this is the standard correlation coefficient (qv).

polythetic: of classification techniques, the opposite of monothetic (qv).

pseudoreplication: the testing of a null hypothesis with an inaccurately inflated estimate of the number of degrees of freedom, a potentially serious class of statistical error covered in Section 1.5.2.

pseudovalues: the multiple estimates generated by a jack-knifing procedure (qv).

ratio data: continuous data (qv) for which ratios may meaningfully be calculated. This includes the majority of continuous data – see also interval data.

regression: any model in which one dependent variable is assumed to be composed of linear combinations of explanatory variables plus unexplained error: $Y_i = A + B \times X1_i + C \times X2_i + \cdots + E_i$, where Y is the dependent variable, A is the intercept (qv), $X1, X2, \ldots$ are the explanatory variables, and E_i is the ith error term.

regression coefficient: the gradient in a regression model: the terms A, B, C in the general regression equation given under 'regression'.

residual: the unexplained error after a model has been fitted to data (Figure 3.5).

Shannon's diversity index: a diversity index defined in Section 2.3.

Shannon–Weaver index: an incorrect alternative name for the Shannon index.

Shannon–Wiener index: an alternative name for the Shannon index.

significant: of a likelihood deemed to be sufficiently low that the null hypothesis may be rejected. Most researchers most of the time use 1:20 (or $p = 0.05$) as their significance value, although this choice is arbitrary. It means that entirely random data will be expected to generate five significant results if subjected to 100 independent inferential tests.

similarity analysis: an alternative name for cluster analysis (qv).

Simpson's diversity index: a diversity index defined in Section 2.2.

Sørensen distance, Sørenson's similarity coefficient: a simple distance function defined in Table 5.1 which only needs presence–absence data.

Spearman's correlation coefficient: a non-parametric version of Pearson's correlation coefficient (qv), calculated by recoding all data as their ranked values (1 = smallest, 2 = second smallest, etc.) then calculating Pearson's correlation coefficient.

standard deviation: a commonly used index of the dispersion of data, defined as the square root of the variance (where variance is the mean value of residuals squared).

transformation: the recoding of a dataset by replacing one variable by a second which is a mathematical function of the first. The most commonly applied transformation is the logarithmic (qv).

UPGMA: a rule for forming clusters (Table 8.5), standing for unweighted pair-groups method using arithmetic averages or UPGMA clustering.

Ward's technique: a Rule for forming clusters, commonly used to produce dendrograms (Section 8.3.5).

weighted mean: a mean calculated on a dataset where some values are repeated (hence given more weight) – see Section 7.2.1.

Bibliography

Anderberg, M.R. (1973). *Cluster Analysis for Applications.* Academic Press, New York.

Anderson, A.J.B. (1971). Ordination methods in ecology. *Journal of Ecology* **59**, 713–726.

Battarbee, R.W. (1984). Diatom analysis and the acidification of lakes. *Phil. Trans. Roy. Soc. B* **305**, 451–477.

Beals, E.W. (1984). Bray–Curtis ordination: an effective strategy for analysis of multivariate ecological data. *Advances in Ecological Research* **14**, 1–55.

Benzécri, J.P. (1992). *Correspondence Analysis Handbook.* Marcel Dekker, New York.

Bobbink, R. (1991). Effects of nutrient enrichment in Dutch chalk grasslands. *Journal of Applied Ecology* **28**, 28–41.

Bobbink, R. and Willems, J.H. (1987). Increasing dominance of Brachypodium pinnatum in chalk grasslands: a threat to a species-rich ecosystem. *Biological Conservation* **40**, 301–314.

Bray, J.R. and Curtis, J.T. (1957). An ordination of the upland forest communities of southern Wisconsin. *Ecological Monographs* **27**, 325–349.

van Breeman, N. and van Dijk, H.F.G. (1988). Ecosystem effects of atmospheric deposition of nitrogen in the Netherlands. *Environmental Pollution* **54**, 249–274.

Brown, K.A. and Roberts, T.M. (1988). Effects of ozone on foliar leaching in Norway spruce: confounding effects due to NOx production during ozone generation. *Environmental Pollution* **55**, 55–73.

Brown, V.K. and Southwood, T.R.E. (1987). Secondary succession: patterns and strategies. In *Colonisation, Succession and Stability* (ed. A.J. Gray, M.J. Crawley and P.J. Edwards), pp. 315–338. 26th Symposium of the British Ecological Society. Blackwell Scientific Publications.

Chardy, P., Glemarec, M. and Laurec, A. (1976). Application of inertia methods to benthic marine ecology: practical implications of the basic options. *Estuarine Coastal Marine Science* **4**, 179–205.

Chapin, F.S., Sala, O.E., Burke, I.C., Grime, J.P., Hooper, D.U., Lauenroth, W.K., Lombard, A., Mooney, H.A., Mosier, A.R., Naeem, S., Pacala, S.W., Roy J. Steffen, W.L. and Tilman, D. (1998). Ecosystem consequences of changing biodiversity: experimental evidence and a research agenda for the future. *BioScience* **48**, 45–52.

Chatfield, C. and Collins, A.J. (1980). *Introduction to Multivariate Statistics.* Chapman and Hall, London and New York.

Cliff, N. (1988). The eigenvalues-greater-than-one rule and the reliability of components. *Psychological Bulletin* **103**, 276–279.

Clifford, H.T. and Stevenson, W. (1975). *An Introduction to Numerical Classification.* Academic Press, New York.

Cormack, R.M. (1971). A review of classification. *Journal of the Royal Statistical Society A* **134**, 321–367.

Crozier, R.H. (1992). Genetic diversity and the agony of choice. *Biological Conservation* **61**, 11–16.

Czekanowski, J. (1913). *Zarys Metod Statystycznyck*, Warsaw.

Daubenmire, R.F. (1966). Vegetation: identification of typal communities. *Science* **151**, 291–298.

De Leuw, J. (1983). On the prehistory of the correspondence analysis. *Statistica Neerlandica* **37**, 161–164.

Digby, P.G.N. and Kempton, R.A. (1987). *Multivariate Methods in Ecology*. Chapman and Hall, London.

Dixit, S.S. (1986). Diatom-inferred pH calibration of lakes near Wawa, Ontario. *Canadian Journal of Botany* **64**, 1129–1133.

Efron, B. (1979). Bootstrap methods: another look at the jack knife. *Annals of Statistics* **7**, 1–26.

Ellin, S.J. (1989). The effect of acid rain and ozone on lichens. Unpublished Ph.D. thesis, Bradford University.

Faith, D.P., Minchin, P.R. and Belbin, L. (1987). Compositional dissimilarity as a robust measure of ecological distance. *Vegetatio* **69**, 57–68.

Fisher, R.A. (1936). The use of multiple measurements in taxonomic problems. *Annals of Eugenics* **7**, 179–188.

Fisher, R.A. (1940). The precision of discriminant functions. *Annals of Eugenics* **10**, 422–429.

Frampton, G.K., van den Vrink, P.J. and Gould, P.J.L. (2000). Effects of spring precipitation on a temperate arable collembolan community analysed using principal response curves. *Applied Soil Ecology* **14**, 231–248.

Gardiner, V. (1979). Estimation of drainage density from topological variables. *Water Resources Research* **15**, 909–917.

Gauch, H.G. (1979). *COMPCLUS – A FORTRAN Program for Rapid Initial Clustering of Large Data Sets*. Cornell University, Ithaca, N.Y.

Gauch, H.G. (1980). Rapid initial clustering of large data sets. *Vegetatio* **42**, 103–111.

Gauch, H.G. and Wentworth, T.R. (1976). Canonical correlation analysis as an ordination technique. *Vegetatio* **33**, 17–22.

Gauch, H.G. and Whittaker, J.B. (1981). Hierarchical classification of community data. *Journal of Ecology* **69**, 135–152.

Gittins, R. (1985). *Canonical Analysis. A Review with Applications in Ecology*. Springer-Verlag, Berlin.

Goodall, D.W. (1953). Objective methods for the classification of vegetation. I. The use of positive interspecific correlation. *Australian Journal of Biology* **1**, 39–63.

Goodall, D.W. (1954a). Vegetational changes and vegetational continua. *Angewachte Pflanzensoziologie* **1**, 168–182.

Goodall, D.W. (1954b). Objective methods for comparing vegetation. III. An essay in the use of factor analysis. *Australian Journal of Botany* **1**, 39–63.

Goodall, D.W. (1978). Numerical classification. In *Classification of Plant Communities* (ed. R.H. Whittaker), pp. 247–286. Junk, The Hague.

Goodman, D. (1975). The theory of diversity-stability relationships in ecology. *Quarterly Review of Biology* **50**, 237–266.

Gower, J.C. (1966). Some distance properties of latent root and vector methods used in multivariate analysis. *Biometrika* **53**, 325–328.

Gower, J.C. (1975). Generalised Procrustes analysis. *Psychometrika* **40**, 33–51.

Green, R.E. (1996). Factors affecting the population density of the corncrake *Crex crex* in Britain and Ireland. *Journal of Applied Ecology* **33**, 237–248.

Grieg-Smith, P. (1964). *Quantitative Plant Ecology*, 2nd edn. Butterworths, London.

Grieg-Smith, P. (1983). *Quantitative Plant Ecology*, 3rd edn. Blackwells Scientific, Oxford.

Hair, J.F. *et al.* (1995). *Multivariate Data Analysis with Readings*, 4th edn. Prentice Hall.

Harris, R.J. (1975). *A Primer of Multivariate Statistics*. Academic Press, New York and London.

Hill, M.O. (1973a). Diversity and evenness: a unifying notation and its consequences. *Ecology* **54**, 427–431.

Hill, M.O. (1973b). Reciprocal averaging: an eigenvector method of ordination. *Journal of Ecology*, **61**, 237–249.

Hill, M.O. (1974). Correspondence analysis: a neglected multivariate method. *Journal of the Royal Statistical Society Series C* **23**, 340–354.

Hill, M.O. (1979a). *DECORANA – A FORTRAN Program for Detrended Correspondence Analysis and Reciprocal Averaging*. Section of Ecology and Systematics, Cornell University, Ithaca, N.Y.

Hill, M.O. (1979b). *TWINSPAN – A FORTRAN Program for Arranging Multivariate Data in an Ordered Two Way Table by Classification of the Individuals and the Attributes*. Section of Ecology and Systematics, Cornell University, Ithaca, N.Y.

Hill, M.O. (1994). *DECORANA 1.0, TWINSPAN 1.0, TABLCORN 1.0 Program Manual*. NERC, Swindon, UK.

Hill, M.O. and Gauch, H.G. (1980). Detrended correspondence analysis: an improved ordination technique. *Vegetatio* **42**, 47–58.

Hill, M.O., Bunce, R.G.H. and Shaw, M.W. (1975). Indicator species analysis, a divisive polythetic method of classification and its application to a survey of native pinewoods in Scotland. *Journal of Ecology* **63**, 597–613.

Hills, M. (1977). Book review. *Applied Statistics* **26**, 339–340.

Hirschfeld, H.O. (1935). A connection between correlation and contingency. *Proceedings of the Cambridge Philosophical Society* **31**, 520–524.

Hopkin, S. (1997). *The Biology of the Springtails*. Oxford University Press, Oxford.

Hotelling, H. (1933). Analysis of a complex of statistical variables into principal components. *Journal of Educational Psychology* **24**, 417–441, 498–520.

Hotelling, H. (1936). Relationships between two sets of variables. *Biometrika* **28**, 321–377.

Hughes, J.W. and Cass, W.B. (1997). Pattern and process of a floodplain forest, Vermont, USA: Predicted responses of vegetation to perturbation. *Journal of Applied Ecology* **34**, 594–612.

Hurlbert, S.H. (1984). Pseudoreplication and the design of ecological field experiments. *Ecological Monographs* **54**, 187–211.

Jaccard, P. (1928). Die Statisch-floristische method als grundlage der pflanzensoziologie. *Aberhalden, Handbuch Biologisch Arbeitsmethod* **11**, 165–202.

Jackson, D.A. (1993). Stopping rules in principal components analysis: a comparison of heuristical and statistical approaches. *Ecology* **74**, 2204–2214.

Janssen, J.G.M. (1975). A simple clustering procedure for the preliminary classification of very large datasets of phytosociological relevées. *Vegetatio* **30**, 67–71.

Jeffers, J.N.R. (1995). The statistician and the computer. *Journal of Applied Statistics* **22**, 227–234.

Jeffers, J.N.R. (1996). Multivariate analysis of a reference collection of elm leaves. *Journal of Applied Statistics* **23**, 571–587.

Jongman, R.H.G., ter Braak, C.J.F. and van Tongeren, O.F.R. (1995). *Data Analysis in Community and Landscape Ecology*. Cambridge University Press, Cambridge.

Jukraine, H.V. and Laioho, R. (1995). Long term effects of water level drawdown on the vegetation of drained pine mires in southern Finland. *Journal of Applied Ecology* **32**, 785–802.

Kaiser, H.F. (1958). The varimax criterion for analytic rotation in factor analysis. *Psychology* **23**, 187–200.

Kenkel, N.C. and Booth, T. (1992). Multivariate analysis in fungal ecology. In *The Fungal Community* (ed. G.C. Carroll and D.T. Wicklow), pp. 209–227. Marcel Dekker Inc., New York.

Kent, M. and Ballard, J. (1988). Trends and problems in the application of classification and ordination methods in plant ecology. *Vegetatio* **78**, 109–124.

Kent, M. and Coker, P. (1992). *Vegetation Description and Analysis, A Practical Approach*. Wiley, London.

Krebs, C.J. (1985). *Ecology: The Experimental Analysis of Distribution and Abundance*. Harper and Row, New York.

Kruskal, J.B. and Wish, M. (1978). *Multidimensional Scaling*. Sage, Beverly Hills, CA.

Laine, J., Vasander, H. and Laiho, R. (1995). Long-term effects of water level drawdown on the vegetation of drained pine mires in southern Finland. *Journal of Applied Ecology* **32**, 785–802.

Lance, G.N. and Williams, W.T. (1967). A general theory of classificatory sorting strategies. 1. Hierarchical systems. *Computer Journal* **11**, 373–380.

Landau, S.I. and Ramson, W.S. (1988). *Chambers English Dictionary*. Chambers, Cambridge.

Lawrence, P.N. (1979). Some observations on the taxonomy and ecology of *Onychiurus armatus* and their wider implications in agriculture and evolution. *Revue d'Ecologie et Biology du Sol* **16**, 259–277.

Louppen, J.W.M. and van der Maarel, E. (1979). CLUSLA: a computer program for the clustering of large phytosociological data sets. *Vegetatio* **40**, 107–114.

McArthur, R.H. (1955). Fluctuations of animal populations and a measure of community stability. *Ecology* **36**, 533–536.

McCune, B. and Mefford, M.J. (1999). *PC-ORD. Multivariate Analysis of Ecological Data, Version 4.01*. MjM Software Design, Gleneden Beach, Oregon.

McDowall, D., McCleary, R., Meidinger, E.E. and Hay, R.A. (1980). *Interrupted Time Series Analysis*. Quantitative Applications in the Social Sciences, Paper 21, Sage, London.

McLeod, A.R., Shaw, P.J.A. and Holland, M.R. (1992). The Liphook forest fumigation project: studies of sulphur dioxide and ozone effects on coniferous trees. *Forest Ecology and Management* **51**, 121–127.

McLeod, A.R. (1995). An open-air system for exposure of young forest trees to sulphur dioxide and ozone. *Plant, Cell and Environment* **18**, 215–226.

Magurran, A.E. (1991). *Ecological Diversity and its Measurement*. Chapman and Hall, London.

Manly, B.J.F. (1986). *Multivariate Methods – A Primer*. Chapman and Hall, London.

Manly, B.J.F. (1991). *Randomization and Monte Carlo Methods in Biology*. Chapman and Hall, London.

Margalef, R. (1958). Information theory in ecology. *General Systems* **3**, 36–71.

Margules, C. (1986). Conservation evaluation in practice. In *Wildlife Conservation Evaluation* (ed. M.B. Usher), Chapman and Hall, London, pp. 298–314.

Margules, C. and Usher, M.B. (1981). Criteria used in assessing wildlife conservation potential: a review. *Biological Conservation* **21**, 79–109.

Mather, P.M. (1976). *Computational Methods of Multivariate Analysis in Physical Geography*. Wiley, London.

Matthew, C., Tillman, R.W., Hedley, M.J. and Thompson, M.C. (1988). Observations on the relationship between soil fertility, pasture botanical composition and pasture growth rate for a North Island lowland pasture. *Proceedings of the New Zealand Grassland Association* **48**, 93–98.

Matthew, C., Lawoko, C.R.O., Korte, C.J. and Smith, D. (1994). Application of canonical discriminant analysis, principal component analysis and canonical correlation analysis as tools for evaluating differences in pasture botanical composition. *New Zealand Journal of Agricultural Research*, **37**, 509–520.

Mielke, P.W., Jr. (1984). Meteorological applications of permutation techniques based on distance functions. In *Handbook of Statistics*, Vol. 4 (ed. P.R. Krishnaiah and P.K. Sen), pp. 813–830. Elsevier.

Minchin, P.R. (1987). An evaluation of the relative robustness of techniques for ecological ordination. *Vegetatio* **69**, 89–107.

MjM Software Design (1995). *Pc Ord: Multivariate Analyses of Environmental Data, Version 2.0*. MjM Software Design, Oregon.

Montgomery, D.C. and Peck, E.A. (1982). *Introduction to Linear Regression Analysis*. Wiley, New York.

Morgan, B.J.T. (1981). Three applications of methods of cluster analysis. *The Statistician* **30**, 205–223.

Morrison, D.F. (1967). *Multivariate Statistical Methods*. McGraw-Hill, New York.

Mueller-Dombois, D. and Ellenberg, H. (1974). *Aims and Methods of Vegetation Ecology*. Wiley, New York.

Nishisato, S. (1980). Analysis of categorical data – Dual scaling and its applications. *Mathematical Expositions* **24**, University of Toronto Press, Toronto.

Økland, R.H. (1999). On the variation explained by ordination and constrained ordination axes. *Journal of Vegetation Science* **10**, 131–136.

Oksanen, J. and Minchin, P.R. (1997). Instability of ordination results under changes in input data order: explanations and remedies. *Journal of Vegetation Science* **8**, 447–454.

Olson, J.S. (1958). Rates of succession and soil changes on southern Lake Michigan sand dunes. *Botanical Gazette* **119**, 125–170.

Orloci, L. (1966). Geometric methods in ecology. I. The theory and application of some ordination methods. *Journal of Ecology* **54**, 193–215.

Orloci, L. (1978). *Multivariate Analysis in Vegetation Research*, 2nd edn. Junk, The Hague.

Parnell, J. and Waldren, S. (1996). Detrended correspondence analysis in the ordination of data for phenetics and cladistics. *Taxon* **45**, 71–84.

Peerson, K. (1901). On lines and planes of closest fit to systems of points in space. *Philosophical Magazine, Sixth Series* **2**, 559–572.

Peet, R.K., Knox, R.G., Case, J.S. and Allen, R.B. (1988). Putting things in order; the advantages of detrended correspondence analysis. *American Naturalist* **131**, 924–934.

Podani, J. (1990). *Syn-tax IV. Computer Programs for Data Analysis in Ecology and Systematics on IBM-PC and Macintosh Computers*. Trieste.

Prentice, I.C. (1977). Non-metric ordination methods in ecology. *Journal of Ecology* **65**, 85–94.

Prentice, I.C. (1980). Vegetation analysis and other invariant gradient models. *Vegetatio* **42**, 27–34.

Rao, V.R. (1952). *Advanced Statistical Methods in Biometric Research*. Wiley, London.

Richman, M.B. (1988). A cautionary note concerning a commonly applied eigenanalysis procedure. *Tellus* **40B**, 50–58.

Roux, G. and Roux, M. (1967). A propos de quelques méthodes de classification en phytosociologie. *Revue de Statistique Appliquée* **15**, 59–72.

Salton, G. and Wong, A. (1978). Generation and search of cluster files. *Association for Computing Machinery Transactions on Database Systems* **3**, 321–346.

Seal, H. (1964). *Multivariate Statistical Analysis for Biologists*. Spottiswood, Ballantyn and Co., London.

Shannon, C. and Weaver, W. (1949). *The Mathematical Theory of Communication*. University of Illinois Press, Urbana.

Shaw, P.J.A. (2000). *The Acid Tests: Studies on the Ecological Effects of Atmospheric Pollution*. Innogy, Swindon.

Shaw, P.J.A. and Lankey, K. (1994). Studies on the Scots pine mycorrhizal fruitbody succession. *The Mycologist* **8**, 172–175.

Shaw, P.J.A., Dighton, J., Poskitt, J. and McLeod, A.R. (1992). The effects of sulphur dioxide and ozone on the mycorrhizas of Scots pine and Norway spruce in a field fumigation system. *Mycological Research* **96**, 785–791.

Simpson, E.H. (1949). Measurement of diversity. *Nature* **163**, 688.

Sneath, P.H.A. and Sokal, R.R. (1973). *Numerical Taxonomy*. Freeman, San Francisco.

Sokal, R.R. (1995). *Biometry: The Principles and Practice of Statistics in Biological Research*, 3rd edn. W.H. Freeman, New York.

Sokal, R.R. and Michener, C.D. (1958). A statistical method for evaluating systematic relationships. *University of Kansas Science Bulletin* **38**, 1409–1438.

Sorensen, T. (1948). A method of establishing groups of equal amplitude in plant sociology based on similarity of species content. *Det Kongelige Danske Videnskabernes Selskab, Biologische Skrifter*, Bind V, nr. 4, Copenhagen.

Southwood, T.R.E. (1978). *Ecological Methods*. Methuen, London.

Spearman, C. (1904). 'General intelligence', objectively determined and measured. *American Journal of Psychology* **15**, 201–293.

SPSS (2000). *SPSS 10 for Windows*. SPSS Inc., Chicago.

Stoermer, E.F., Wolin, J.A. and Schelske, C.L. (1993). Paleolimnological comparison of the Laurentian great lakes based on diatoms. *Limnology and Oceanography* **38**, 1311–1316.

Stubberfield, L.C.F. and Shaw, P.J.A. (1990). A comparison of tetrazolium reduction and FDA hydrolysis with other measures of microbial activity. *Journal of Microbiological Methods* **12**, 151–162.

Swain, P.H. (1978). Fundamentals of pattern recognition in remote sensing. In *Remote Sensing: the Quantitative Approach*. McGraw-Hill, New York.

Swann, B.B. (1953). Statistical computation and electronic machines. *The Incorporated Statistician* **4**, 81–91.

ter Braak, C.J.F. (1986). Canonical correspondence analysis: a new eigenvector technique for multivariate direct gradient analysis. *Ecology* **67**, 1167–1179.

ter Braak, C.J.F. (1987). *CANOCO. A FORTRAN Program for Canonical Community Ordination by [Partial] [Detrended] [Canonical] Correspondence Analysis (Version 2.0)*. TNO Institute of Applied Computer Science, Wageningen, Netherlands.

ter Braak, C.J.F. (1988). CANOCO: an extension of DECORANA to analyse species–environment relationships. *Vegetatio* **75**, 159–160.

ter Braak, C.J.F. (1990). *Update notes: CANOCO Version 3.1*. Agricultural mathematics group, Wageningen, Netherlands.

Tilman, D., Wedin, D. and Knops, J. (1996). Productivity and sustainability influenced by biodiversity in grassland ecosystems. *Nature* **379**, 718–720.

Usher, M.B. (1975). Natural communities of plants and animals in disused quarries. *Journal of Environmental Management* **8**, 223–236.

Van den Brink, P.J. and ter Braak, C.J.F. (1997). Ordination of responses to toxic stress in experimental ecosystems. *Toxicology and Ecotoxicology News* **4**, 173–177.

Van den Brink, P.J. and ter Braak, C.J.F. (1998). Multivariate analysis of stress in experimental ecosystems by principal response curves and similarity analysis. *Aquatic Ecology* **32**, 163–178.

Van den Brink, P.J. and ter Braak, C.J.F. (1999). Principal response curves: analysis of time-dependent multivariate responses of biological communities to stress. *Environmental Toxicology and Chemistry* **18**, 138–148.

van Tongeren, O. (1986). FLEXCLUS, an interactive program for the classification and tabulation of data. *Acta Botanica Neerlandica* **35**, 137–142.

Vankat, J.L. (1982). A gradient perspective on the vegetation of Sequoia National Park. *Madrono* **29**, 200–214.

Vickers, A. (2001). *The Evaluation of Woodland Status by means of Botanical Indicator Species*. PhD thesis, Sheffield Hallam University.

Ward, J.H. (1963). Hierarchical grouping to optimise an objective function. *American Statistical Association Journal* **58**, 236–244.

Wartenberg, D., Ferson, S. and Rohlf, F.J. (1987). Putting things in order; a critique of detrended correspondence analysis. *American Naturalist* **129**, 434–448.

Whittaker, R.H. (1956). Vegetation of the Great Smoky Mountains. *Ecological Monographs* **23**, 41–78.

Wildi, O. (1979). GRID – A space density analysis for recognition of noda in vegetation samples. *Vegetatio* **41**, 95–100.

Wildi, O. (1980). Management and multivariate analysis of large data sets in vegetation research. *Vegetatio* **42**, 175–180.

Williams, W.T. and Lambert, J.M. (1959). Multivariate methods in plant ecology. I. Association in plant communities. *Journal of Ecology* **47**, 83–101.

Williams, W.T. and Lambert, J.M. (1960). Multivariate methods in plant ecology. II. The use of an electronic digital computer for association analysis. *Journal of Ecology* **48**, 689–710.

Williams, W.T. and Lambert, J.M. (1961). Multivariate methods in plant ecology. III. Inverse association analysis. *Journal of Ecology* **49**, 717–729.

Williams, W.T., Lambert, J.M. and Lance, G.N. (1966). Multivariate methods in plant ecology. V. Similarity analysis and information analysis. *Journal of Ecology* **54**, 427–445.

Wishart, D. (1969). An algorithm for hierarchical classifications. *Biometrics* **25**, 165–170.

Wishart, D. (1987). *CLUSTAN User manual (CLUSTAN 3)*, 4th edn. Computing Laboratory, University of St Andrews, Scotland.

Zimmerman, G.M., Goetz, H. and Mielke, Jr. P.W. (1985). Use of an improved statistical method for group comparisons to study effects of prairie fire. *Ecology* **66**, 606–611.

Index